THE ROMANCE OF
CAPE MOUNTAIN PASSES

Published to mark the 100th anniversary
of the founding of the
South African Institution of Civil Engineering
1903–2003

1903 Cape Society of Civil Engineers
1910 South African Society of Civil Engineers
1948 South African Institution of Civil Engineers
1994 South African Institution of Civil Engineering

The publication of this book has been made possible
through the efforts of the triumvirate consisting of
Brenda Sudano, Ross Parry-Davies and Graham Ross.

THE ROMANCE OF CAPE MOUNTAIN PASSES

GRAHAM ROSS

David Philip
Cape Town

First published 2002 by David Philip Publishers,
an imprint of New Africa Books (Pty) Ltd,
99 Garfield Road, Kenilworth, Cape Town

© 2002 Graham Ross

Second impression 2003

ISBN 0-86486-596-1

All rights reserved

Front cover photo of Gamkaskloof: Obie Oberholzer
Design and layout: Abdul Amien
Reproduction: Castle Graphics, Cape Town
Printing: ABC Press, Cape Town

CONTENTS

Foreword by the Minister of Transport vii
Foreword by the Western Cape Minister of Transport viii
Preface by Patricia Storrar ix
Introduction x
Acknowledgements xiii
General Map xiv–xv

1. THE ROODEZAND PASSES 1
2. PIQUINIERS KLOOF, GREY'S PASS & PIEKENIERSKLOOF PASS 8
3. GANTOUW, HOTTENTOTS HOLLAND KLOOF & SIR LOWRY'S PASS 14
4. MOSTERTSHOEK & MICHELL'S PASS 24
5. ATTAQUAS KLOOF & ROBINSON PASSES 32
6. PLATTEKLOOF, TRADOUW & GARCIA'S PASSES 37
7. KAAIMANSGAT & DUIWELSKOP PASSES 45
8. PAARDEKOP & PRINCE ALFRED'S PASSES 54
9. HEX RIVER PASS & POORT 60
10. CRADOCK KLOOF & MONTAGU PASSES 65
11. CATS PASS & FRANSCHHOEK PASS 71
12. THE HOUW HOEK PASSES 76
13. BAIN'S KLOOF PASS 81
14. MEIRINGSPOORT 89
15. SEWEWEEKSPOORT & BOSLUISKLOOF PASS 96
16. THE MESSELPAD PASS 101
17. COGMANS KLOOF PASS 108
18. KOO OR BURGER'S PASS 114
19. GROOT RIVER & BLOUKRANS PASSES 117
20. THE PASSES ROAD 123
21. SWARTBERG PASS 131
22. PENHOEK PASS 136
23. HUIS RIVER PASS 141

24. Chapman's Peak Drive & the Mountain Roads to Hout Bay	147
25. Boesmanskloof Pass & Buffelspoort	154
26. The Steenbras Mountain Road	159
27. Du Toit's Kloof Pass	164
28. Great Brak Pass	170
29. Outeniqua Pass	176
30. Anenous Pass	183
31. The Gamkaskloof Passes	189
32. What of our Mountain Passes in the Future?	195
Appendix: List of Cape Mountain Passes	199
Notes	210
Bibliography	216
Index	221

FOREWORD
by Minister A M Omar

A journey of a thousand miles begins with a single step.
— Chinese proverb

The author of this book needs to be congratulated on his comprehensive work on Cape mountain passes. His skilful compilation, with the clever use of humour and anecdotes from yesteryear, will appeal as much to engineers as it will to historians and readers with an interest in the landscape and development of the Cape.

The book is also an important record of some of the engineering achievements up to the 1990s. If we take into account the progress made in civil engineering know-how, as well as the sophisticated and large-scale capital equipment currently at our disposal in road-building, the sheer determination and arduous physical effort of the early road-builders are to be admired.

The art and science of road-building is not simply the creation of a means of travelling between two points. Roads are the main arteries of a country, essential for the transportation of people, services and goods. The economic significance of good roads and an effective infrastructure cannot therefore be overemphasised. Thanks to these roads, the Cape has been able to play its vital role in the socio-economic development of South Africa.

As an educational tool, *The Romance of Cape Mountain Passes* is essential reading for civil engineering students. Let us hope that this book will be an encouragement to our future engineers to make their own history in road-building and, in time to come, be an inspiration while they recount their own achievements with pride.

Dr Abdulah M Omar
Minister of Transport
September 2002

FOREWORD
by Minister T Essop

I must admit that even I, as the recently appointed Provincial Minister of Transport, Public Works and Property Management for the Western Cape, have limited knowledge of the history of construction of the magnificent mountain passes of the Province, other than an impression gained from those that I have had the privilege of driving through. This book will bring me up to date, and I am honoured to be given the opportunity to write a foreword to a work that will, no doubt, be owned with pride by roads engineers and others who have a love for history, for roads, mountains and the environment.

I believe that the book is very appropriately titled 'The Romance of Cape Mountain Passes' since I have listened to some of the 'younger' engineers, now veterans themselves, who have been involved in the construction or reconstruction of several of these passes in more recent times. They speak with love for and pride in these projects, in a humble manner, not boastfully. It is obvious to me that the construction of a mountain pass is the quest for a compromise and the achievement of a negotiated partnership between nature, machine and humankind. And teamwork wins the day on each occasion. The synergy of the engineers, surveyors, soils technicians and construction teams, including the large labour components, each member of which owns the project, contributes to this success and enables them to hold the pass with pride as a personal achievement.

The author refers to Mrs Bain moving temporarily into a little cottage at the foot of the Piquiniers Pass, and goes on to say that she 'like a good *padmaker*'s wife, throughout their married life, followed her husband from place to place as the dictates of his work demanded'. While her devotion is admirable and commendable, one should also not forget the wives of the migrant labourers and convicts who were forbidden to accompany their husbands. These women singlehandedly raised their families through subsistence farming while looking forward to the once-a-year visit of their husbands from far away. It is shameful that this situation is still a reality today. On the other hand I am encouraged by the fact that more and more women are today engaged in road design and construction.

I salute these men and women and what they have achieved in providing us with the mountain passes of the Cape and of our country. A special thought of course to the largely unsung contribution of those engaged in hard labour in the construction of the passes. Some part of their spirits must surely live on in the brooding cliffs, the rocky streams, and the winding roads they have produced.

Read this book with interest and with pride, and sense the latent energy of those involved when next you traverse one of these passes.

Minister T Essop
Western Cape MEC for Transport, Public Works and Property Management

PREFACE

This remarkable book deserves the honour of having been published to mark the centenary of the founding of the South African Institution of Civil Engineering. It is, as one would expect, perfectly accurate: every figure, date, level, gradient and destination quoted, from the table of contents to the appendix. But this is no dry-as-dust account of figures and names. From the moment one reads the title, *The Romance of Cape Mountain Passes* (not 'The History of the Cape Mountain Passes'), one realises that this is an engineering book with a difference. It is extremely readable, written in clear, smooth-flowing prose, lit here and there by bright little shafts of humour.

It was early in the nineteenth century, as the technological and industrial revolutions got under way, that the great road- and bridge-builders resumed the work that the Roman engineers had started many centuries previously. In South Africa, opening up its vast elevated plateau entailed penetrating it by means of passes – most of them rising through its peripheral mountain girdle. Every pioneer civil engineer was perforce a topographer who had to survey and map a possible route on horseback, carrying his instruments on a mule-wagon and often having to climb on foot to the highest points to establish a trigonometrical network to work from. For weeks they camped out under canvas in all kinds of weather, calculating and plotting their work, often under appalling conditions; and attended to the feeding, housing and general well-being of the labour force as well as ordering and fetching material and supplies.

Thomas Bain's mapping of Prince Alfred's Pass is a beautiful example of neat and accurate draughting. Civil engineers had to be competent draughtsmen. Thomas ('the man with the theodolite eye', as he was known to his colleagues) did not have the technology available to the modern engineer – no maps, no modern explosives, no earth-moving machinery, no water tankers or concrete mixers, no quarry equipment, no power drills. Any engineering contractor, if approached today to build some kilometres of road, would not complete the undertaking without having all of these in place plus a few million rands, possibly a helicopter and of course many computers. When I told Andrew Roberts of Murray & Roberts Holdings what it had cost to build the Swartberg Pass, he exclaimed, 'Good heavens! I couldn't build half a kilometre of tarred road for that today.'

Whenever we drive through any mountain passes, let us remember the many men who made our smooth and comfortable progress possible amid such majestic surroundings. Dr Ross has traced the long and arduous saga of the lonely pass-builders contending with almost insuperable obstacles – from the days of pick, shovel and gunpowder to today's highly mechanised and computerised operations and high explosives. And for this literary labour of love, we are grateful to him for so meticulously putting to paper the fascinating details and spirit of these men's heroic accomplishments – to inform and delight both lay and professional readers as well as being a rare record for posterity of the engineer's perennial contest with nature.

Patricia Storrar, author of *Colossus of Roads*

INTRODUCTION

The reasons for writing this book are twofold. Firstly, I feel that there is far too little record of the history of civil engineering works in our country. And as Winston Churchill is alleged to have said 'the further backward you look, the further forward you can see', such a record should have merit. I decided to see if I could do something about it, seeing that I am now retired and supposedly my own master.

So I got stuck in, researching and recording the various sources which are available here. I started on the Cape, as that is where the major portion of my engineering has been done – and I soon realised that I must in fact restrict myself to the Cape: my available time is limited, and the Cape offers sufficient challenge to keep me on my toes. The result is a 400-plus-page production, *Mountain passes, roads and transportation in the Cape: A research document*. This consists of an index of over 490 Cape passes (reproduced as an appendix to this work), a list of major construction dates, and a chronology recording data extracted from the sources which, in turn, are detailed in an annotated bibliography.

The second reason for writing this book is that the flow of data for inclusion in my research document was slowing, and more and more people were harassing me to write something about the mountain passes I had researched. I had compiled, for the South African Institution of Civil Engineering, 18 articles in a series called *Reminiscences about Cape mountain passes*. These had been published over a period in the Institution's magazine, *Civil Engineering*, and were then offered as a collection in 1998. They were stories, tales and recollections of specific projects or happenings which I had bullied friends into contributing. I thought that it would be nice to partner this series with another, which would record the histories of various Cape mountain passes, using for this purpose some of the data I had collected in my research document.

I opted to tackle the passes (more or less) in the chronological sequence in which they were discovered or pioneered, in the recorded history of our country. The result would be that the coverage would gradually spread out from Cape Town, following the development of (road) mountain passes as the colonial settlement pushed its way into the interior. This appeared to me to offer a logical progression for the chapters.

The history of each pass, as its turn came, would be followed from that early date through to the present time in one chapter – the narrative would not jump from pass to pass so as to keep a chronological sequence of events.

As far as coverage is concerned, it was obvious that only some of the 490-odd passes could be included, but what was not so easy to decide was what to cover about these passes.

First of all, the book is a record of the various stages in the histories of the passes about which I have written. Next, I had decided at the outset that this was not to be theoretical, or scientific, or even very professionally an engineering book. You will generally not find schedules of quantities, geometric details, structural design information or materials data, although a mention of radius, gradient, depth of cut or height of fill may have crept in here and there. There is even, if I remember correctly, a mention of the thickness of the base-course on one occasion – old habits die hard! But any such mentions will be purely in passing, and they will not interfere too much with the main story line.

So these were the guidelines which I attempted to follow.

✦✦✦✦✦✦✦✦✦

It is probably a good idea to define a mountain pass so that we all know what we are talking about. The term 'mountain pass' as used in this work is taken to include poorts that give access from one side of a mountain to another, such as Meiringspoort. And passes do not necessarily have to go up, over a mountain and then down again like Swartberg Pass; they can also go down to a river and then up the other side, as Homtini Gorge does; or they can just go down (or up, depending on which way you are travelling), like the Hex River Pass.

Jose Burman, in his monumental book *So high the road* (1963), said

> No two Cape mountain passes are alike,
> either in appearance or evolution;
> their stories are human documents,
> pages in the saga of a young country's development;
> monuments to the vision and labour of great men.

I only hope that the following chapters may help to illustrate the truth of his stirring words.

ACKNOWLEDGEMENTS

I should like to acknowledge the back-up provided by the two other members of the engineering triumvirate, Brenda Sudano and Ross Parry-Davies, as well as the guidance and assistance received from my editor, Russell Martin. Without their combined input, the publication deadline to mark the SAICE Centenary Year would not have been met.

Then, not only did the sponsorship received from the South African National Roads Agency and the Provincial Administration of the Western Cape make it possible to reduce the selling price of the book and to go for a higher standard of production, but their support and belief in the project also provided much appreciated encouragement when doubts crept in or obstacles and difficulties appeared most daunting.

I must also mention the many friends, colleagues, archivists, librarians and fellow researchers who have contributed to the gathering of the data on which this book is based, and who have provided some of the illustrations used in this publication.

In fact, without you, the book would probably not have been written at all.

Graham Ross
Somerset West

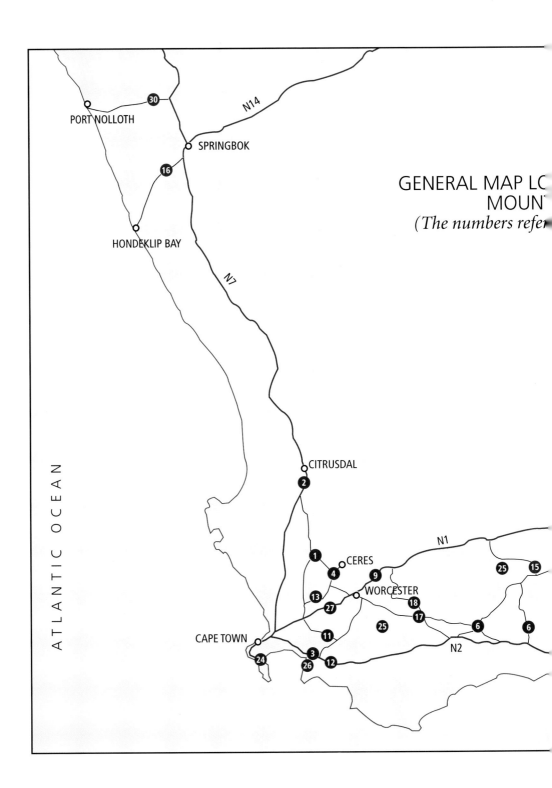

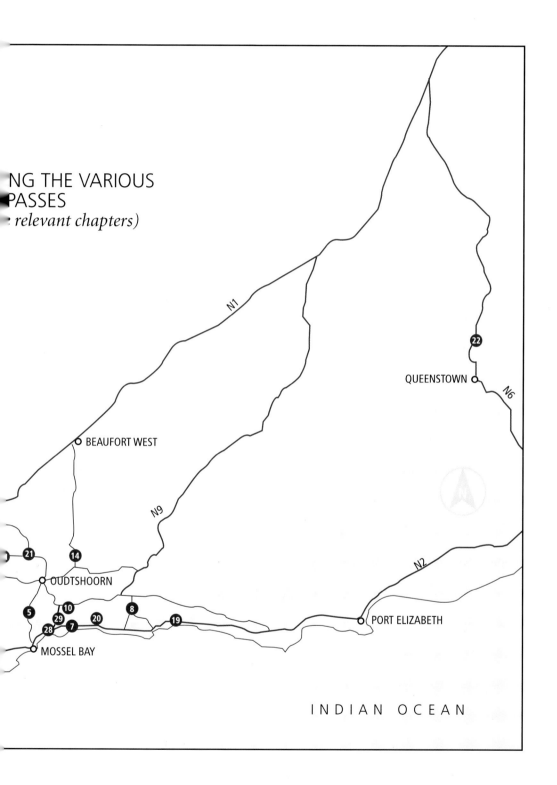

CHAPTER ONE

THE ROODEZAND PASSES

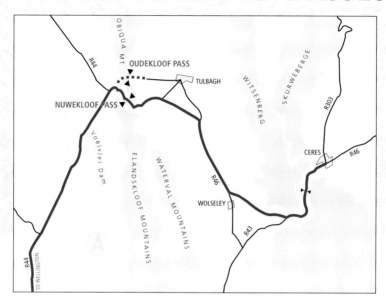

From the earliest days of white settlement at the Cape, the 'Mountains of Africa', as Commander van Riebeeck referred to them, presented a barrier to exploration and trade in the interior. Stretching from the sea at Gordon's Bay they advance behind (to the east of) present-day Paarl, Wellington, Gouda and Porterville and continue further north along the Sandveld. Initially there were only three known places where this barrier could be penetrated: Gantouw or Hottentots Holland Kloof in the south, Piquinierskloof about 170 kilometres to the north, and the Roodezand Passes more or less halfway between them.

I use the plural advisedly, for at various times there were four different passes in use to cross the Mountains of Africa to the Roodezand, as the Tulbagh Valley – 90 kilometres northeast of Cape Town – was known in earlier days. Each pass in turn gave access to the Roodezand, and so each in turn was known as 'the Roodezand Pass'. This can sometimes lead to understandable confusion when researching their histories.[1]

OUDEKLOOF PASS

It was Pieter Potter who discovered the first Roodezand Pass, in 1658. He was the surveyor attached to the cattle-buying expedition sent by Van Riebeeck to barter for cattle with the indigenous Khoi herders. Led by Sergeant Jan van Harwarden, the main body was held up south of Voëlvlei, and Potter took a party of five with instructions to find a way over the mountains in an effort to locate the herds of the local Khoikhoi. He scaled the mountain behind Voëlvlei but the high-level valley there was devoid of stock. He descended again, and pressed northwards until he came to what is now known as Tulbagh Kloof where the Klein Berg River emerges from the mountains on its way to join the Berg River on its journey to the sea.

Potter tried to follow the river upstream into the gorge, scrambling along the bank, but this became increasingly precipitous, so he gave up and returned downstream until he was able to scale a spur on the north side. From a short distance beyond the summit he had a grand view of the Roodezand Valley, and could appreciate that he had found a way through the Mountains of Africa. However, there was no sign of any cattle or sheep so he returned to the main body of the expedition.[2]

Nothing was done to develop this pass for about forty years. When Willem Adriaan van der Stel took over as Governor in 1699, he opened the fertile Roodezand Valley to farming and named it Land of Waveren, after a place in Holland. However, the local settlers continued to refer to the Roodezand, thus named because of the conspicuous red sandstone cliffs in the northwestern corner of the valley. And the pass which was developed over a low neck in the Obiqua Mountain about four kilometres north of Gouda to give access to the area, was known as Roodezand Pass.

It was rather a low-standard pass – though they all were at that time. It had a particularly steep slope on the eastern side. Peter Kolbe wrote in 1731: 'The wagons that pass between this colony [Roodezand] and the Cape ... are generally unloaded at the foot of the mountain and taken to pieces and they and their goods carried over in small parcels on the backs of the cattle and the drivers ... the road across the mountains is very narrow and in many places thick set with trees on both sides.'[3]

In 1748 the local farmers made an effort to improve the pass, but there was a limit to what could be done.[4]

NIEUWEKLOOF PASS

A year or so later it was decided that the only solution was to try a different route, and the farmers were mobilised by one Jacobus du Toit. With much effort they succeeded in opening up a rather basic route through Tulbagh Kloof, on the eastern side, along the right bank of the Klein Berg River. This route had no steep section like the old route, and by the 1760s it had to all intents and purposes superseded the original pass. Seeing that it gave access to Roodezand, it was of course referred to as Roodezand Kloof! To avoid confusion the original pass was known as the Oude Roodezand Kloof while the new route became the Nieuwe Roodezand Kloof, but these names were soon abbreviated to Oudekloof and Nieuwekloof.[5]

Carl Thunberg, a visiting botanist, has left us a record of his passage through the kloof in June 1772. He came from Saldanha

> to the Little Mountain River, and farther through the Roodezand Kloof (Red Sand Valley) to Wafersland or Roodezand. The cleft through which we passed from the sandy plain that lies towards the Cape, but gradually rises until it comes to Roodezand, is one of the few chasms left by the long range of mountains through which it is possible for a waggon

to pass, though possibly not entirely without danger. In some places it was so narrow that two waggons could not pass each other. At such narrow passes as this it is usual for the drivers to give several terribly loud smacks with their long whip which are heard at a distance of several miles, so that the waggon which arrives first may get through unimpeded before another enters it.[6]

Nieuwekloof in 1811 (William Burchell)

The traveller William Burchell went through a bit later, in 1811. He described Nieuwekloof as

a narrow winding defile of about three miles in length, just enough to allow a passage for the Little Berg River on each side of which the mountains rise up abrupt and lofty. Their rocky sides are thickly clothed with bushes and trees from their very summits down to the water, presenting a beautiful romantic picture adorned with their variety of foliage. Along the steep and winding sides a road has been cut, which follows the course of the river at a height above it generally between 50 and 100 feet, in one part rising much higher and in another descending to the bottom and leading through the river, which at this time [April] was not more than three feet deep, although often so swollen by the rains as to be for a day or two quite impassable.[7]

These drifts used to hold up traffic. In 1776, Andrew Steedman had had his wagon sink into the sand of a drift 'and it was not without extreme difficulty that we succeeded in extricating it'.[8]

There are intriguing references to tolls on this road. The historian Theal records that on 28 February 1807 a turnpike was established on Roodezand Kloof to levy the following tolls: 'a chaise, 4 shillings; loaded waggon, 4 shillings; unloaded waggon, 2 shillings; cart, 2 shillings; saddle horse, 1 shilling; every 20 oxen or cows, 4 shillings; every 100 sheep, 4 shillings'.[9] This information agrees with Burchell's statement that in 1811 he paid the 'trifling toll which is levied to defray the expences of keeping this pass in repairs' at the Tulbagh entrance. But the traveller Henry Lichtenstein mentions paying in 1803 'a certain toll for each wagon, which goes to the overseer appointed to keep the road in repair'. Was the overseer levying a private, unofficial toll for his personal benefit in 1803? Incidentally, both Lichtenstein and Burchell were extremely critical of the poor state of repair of the pass road. Burchell also mentions that to avoid paying the toll on Nieuwekloof, cattle drovers made use of Oudekloof.

In 1805 Roodezand was officially named Tulbagh, and a landdrost was appointed there. After this many people started referring to Nieuwekloof as 'Tulbagh Kloof'.[10] Charles Michell, Surveyor-General, Chief Inspector of Works and Civil Engineer, in a landmark paper to the Royal Geographical Society in 1836, described Tulbagh Kloof as 'a natural gap formed by the passage which the Klein Berg River has made for itself ... This pass, though rugged, offers no serious obstacle; the ascents and descents are, for this country, scarcely worth noticing'.[11]

Shortly thereafter, as part of the energetic pass-building programme under Colonial Secretary John Montagu, Michell's engineers began looking at ways in which the road through Nieuwekloof could be improved.

Thomas Bain's Nieuwekloof Pass (Transnet Heritage Library)

TULBAGH KLOOF PASS

Thomas Bain inspected the kloof in 1855, and recommended a route on the western side, along the left bank of the Klein Berg River. He was thus able to cut out the two drifts which had occasioned such delay and frustration. The new route was built during 1859 and 1860 by the Divisional Council, along the line set out by Bain.[12] This Roodezand/Nieuwekloof/Tulbagh Kloof Pass carried all the road traffic through the kloof for over a hundred years. True, it picked up a bituminous surface along the way, but the road remained basically unchanged.

When Bain was transferred to build a railway line through the kloof in 1873 and 1874, he built it on the same side of the river as his road. This left bank was getting a little crowded in some places, and when in 1935 it was necessary to widen Bain's road to a total formation width of 21 feet it required special efforts in these places. The toe of the railway embankment was pushed back a few feet in one section by cutting back in six-foot sections and installing a retaining wall, while elsewhere the parapet wall on the river side was removed and replaced by an unusual cantilevered guard rail. The road through the kloof was successfully widened.[13]

NUWEKLOOF PASS

By the time this roadway had been in use for a century, a 21-foot formation was no longer sufficient to accommodate the traffic safely and efficiently. Because there was insufficient space on the west side to carry out construction while still accommodating traffic, the route had to be carried on the east side, back on the right bank of the Klein Berg River. The necessary expertise and equipment were now available to do the rock cuts needed to construct a road whose geometric design (flatter curves and increased sight distances) would cater for today's larger and faster vehicles. Of course, with reinforced concrete it was also much easier than it would have been in Bain's day, to build the two bridges at the river crossings.

The Provincial Roads Department swung into action, topographic and materials surveys were carried out, a line chosen, cross-sections and quantities determined, and a new road to the desired standard was constructed. This road was opened in 1968. The cartographers meantime had decided to label the road through the Klein Berg River Kloof 'Nuwekloof Pass' once again, with the spelling updated.

When you travel through the pass you will be able to see the 1860 road and the railway line on the other side of the river. You may also park at some safe place and walk onto and along this early construction. Allow sufficient time to pause along the way to absorb the full details of that early construction.

You can also look up across the river, and appreciate the 1968 construction, because it is not only the old projects which are deserving of appreciation and admiration.

✦✦✦✦✦✦✦✦✦✦

This is probably a good place to review the ways in which the skill and science of road-building progressed during the nineteenth and twentieth centuries. From the early days we

'Passing a kloof' (Samuel Daniell)

find two aspects: the road authorities, who were responsible for authorising and finding the finance for road construction, and the road-builders or *padmakers*, who actually designed and built the roads in the field. The histories of these two contributors are of necessity interwoven, as we can see from the historical record.

As early as 1805 Ordinance 164 required that 'Landdrost and Heemraden are to look to the making and keeping in repair, of Streets and Roads; and they are in particular to endeavour, that all the Passes over the Mountains or Rivers, by which the Produce of the Colony is to be conveyed, either to Cape Town or to any other market, are put in the best possible state'. To help them in their maintenance duties in the remoter country districts, Ordinance 273 required that 'the Fieldcornets, each in their respective Districts, shall pay attention to the improvement and repairs of the Public Roads'.[14]

In 1806, because of the bad state of the roads in the country districts, landdrosts were authorised to appoint an Overseer of Roads for each district, who was empowered to call upon the inhabitants to furnish a proportion of their slaves for the repair of the roads.[15] So there we have the road authorities – and we see that the actual work was done by slaves 'borrowed' from the local inhabitants. This worked quite well until the emancipation of slaves in the British Empire in 1834. Even though slaves in the Cape were indentured for the succeeding four years, this resulted in the previous roadwork measures becoming ineffective because of the sudden shortage of labour. Appreciate that there were roughly 30,000 slaves in the Cape, out of a total recorded population of 77,000.[16]

In 1843 John Montagu was transferred from Tasmania, where he had gained good experience in the handling and treatment of convicts and their employment on public works, to hold the office of Colonial Secretary at the Cape. He appreciated the need for good roads in the Colony. With the encouragement and backing of the Governor, he created the Central Road Board, which provided a much-needed coordinating road authority. Next he largely saved the road-labour situation with his Ordinance 8 of 1843, which *inter alia* gave the Board powers to elicit convict labour for roadworks. Charles Michell, Andrew Geddes Bain, his son Thomas Bain, Adam de Smidt, Patrick Fletcher and others largely – but not entirely by any means – used convict labour for road construction and repair in the Cape.

John Montagu

In 1855 Sir George Grey was one of those instrumental in forming a new rural administrative body, the Divisional Council, which improved on and combined the best features of the two previous systems – the Landdrost and Heemraden, and the Road Board. Between them the Cape government and the Divisional Councils shared responsibility for road construction and maintenance. For probably the first time, it could be said that road maintenance received a modicum of continuing attention.

After Union in 1910, the Cape retained the Divisional Councils, which had been shown to be the near-ideal local road authorities, and which worked extremely well in conjunction with the Provincial Roads Department: until 1994 in fact. In the other provinces of the Union, the road responsibility remained largely in provincial hands.

Another major player entered the arena and took an interest in the more important interurban routes when the National Roads Act was promulgated in 1935 and the South African National Roads Board came into being.

The degree of responsibility between all these authorities has varied over the years as people saw where improvements could be made. As time passed, the use of convicts on roadworks gradually ceased, and more and more use was made of 'free' labour – which of course was not free as convict labour had been but cost money.

The work was done by a dedicated band of road engineers and other *padmakers*, sometimes departmentally and, over time, more and more by consulting engineers and contractors. But by whomever the work was planned, designed and constructed the team spirit displayed throughout the various units has always been outstanding, especially in the Cape!

CHAPTER TWO

PIQUINIERSKLOOF, GREY'S PASS & PIEKENIERSKLOOF PASS

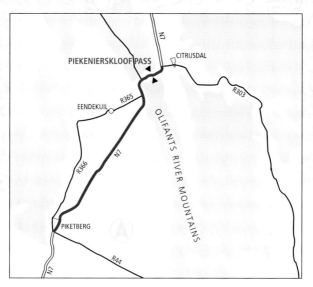

G rey's Pass, now known as Piekenierskloof Pass – situated between Piketberg and Citrusdal about 150 kilometres north of Cape Town on the road to Namibia – is a most important crossing of the barrier formed by the Olifants River mountain chain. The road traverses the mountains at a height of 520 metres above sea level and forms an important link between the Western Cape and the agricultural and mineral-rich areas to the north.

Things seem to go by centuries here. It is known that there was a bridle path through the mountains in 1660, Grey's Pass was opened in 1858, and the modern Piekenierskloof Pass was opened in 1958.

Whence the names? Apparently a contingent of pikemen (*piquiniers*) crossed the mountains here in 1675 in pursuit of a band of Khoi: of course they did not catch their nimble quarry. Thomas Bain's pass was named after Sir George Grey, Governor of the Cape Colony from 1854 to 1859. Upon completion of the present pass the older name was revived, but with the spelling updated.

✦✦✦✦✦✦✦✦✦✦

The stories of the exploring expeditions passing over or round the Olifants River Mountains, in their initial endeavours to find the cattle herds of the Khoi so as to barter for meat for the ships calling at Table Bay, and their subsequent struggles to reach the Namaqualand copper mountains, are totally enthralling. But you will have to look elsewhere for these details: here we must, I fear, concentrate on the pass.[1]

Jan Dankaert led the first expedition across, in 1660. He was fortunate in that the local people showed him a track by which he reached the top, and descended on the other side. Other expeditions followed in his footsteps, and in 1662 Pieter Cruythoff led the first expedition to include a wagon. However, they had to unload the wagon, drag it up the pass (they only had six oxen), reload it to cross the plateau, and then unload again to get it down the other side, once again carrying all their goods and chattels. When they got to the bottom they found the riparian growth so dense that they went on without the wagon, which they buried, using the oxen as pack animals. On their return they did not stop to pick up the wagon – they probably couldn't face the thought of all that pulley-hauley up the mountain. Sergeant de la Guerre had a similar experience in 1663, with the added frustration that when his party returned to pick up their buried wagon they found that the local Khoi had burnt it to get the iron parts.[2]

It was as a direct result of the great difficulties experienced in crossing mountains – and even hills – with wagons that the visiting Commissioner Ryklof van Goens in 1682 issued suitable instructions. In future all wagons were to be constructed such that they could be taken to pieces and transported over the mountains on the backs of the long-suffering oxen. A really brilliant edict, which greatly extended the possible range of exploring and trading expeditions.[3]

Many parties crossed these mountains in the succeeding years. This is part of the description which Lichtenstein left of his 1803 crossing:

> From the steepness of the hill it was impossible to carry the road directly over it, but it forms a zigzag turning repeatedly, though always ascending. At such moments the whole team of oxen cannot be made to draw as one, and the wagon is in danger either of running back, or if it turns too sharp, of being wedged against the rock. We sent our baggage forward very early in the morning, following ourselves an hour later. When about halfway up the kloof we found one of our wagons stuck fast, nor was there any other means for its release but to unload it, and set it right by the exertions of our own strength … The rugged wildness of these lofty regions, the gigantic masses of naked rock, the tremendous height from which one looks down upon the precipices below, makes it almost incomprehensible how a heavy-loaded wagon should ever reach the summit … [4]

To me, this sounded like someone trying to take a wagon up a zigzag footpath. So I was very pleased to read that when Jose Burman traced the old pass up the mountainside to the west, he found only one zig in the road, which also agreed with the (rather basic) description left by Dankaert.[5] I have come to the conclusion that Lichtenstein must have been including as zigzags the occasions when the road turned in narrow kloofs – or possibly something was lost in the translation from the original German.

By the mid-1800s the British Governors at the Cape were encouraging the building of mountain passes to open up communication with the hinterland. It was felt that an up-to-standard pass should be provided over the Olifants River Mountains, but for a long time the

Grey's Pass and the cottage in which the Bain family lived (Gunther Komnick)

Central Road Board dithered between Piquiniers Kloof and Kardouw Pass, 20 kilometres further south.[6]

Finally it was decided that Thomas Bain should build an engineered pass at Piquiniers Kloof. Thomas Bain was born in 1830 in Graaff-Reinet. He served a six-years' apprenticeship under his father, Andrew, mainly on Michell's Pass and Bain's Kloof. In 1854 he was appointed Inspector of Roads for the Western Province, after having passed the prescribed government examination in civil engineering.

Thomas attained an enviable reputation as a locator, designer, builder and supervisor of the construction of mountain passes in the Cape. He was also for a short time a District Engineer in the Railway Department as well as Irrigation and Geological Surveyor of the Colony. A crowning achievement was when he became an Associate Member of the Institute of Civil Engineers in 1877.

The Piquiniers Kloof construction was Thomas Bain's first job where he was himself directly responsible for the construction work – what we would today term the Resident Engineer. Thomas and his wife Johanna moved into a little cottage at the foot of the pass, on the Piketberg side. Johanna, like a good *padmaker*'s wife, followed her husband throughout their married life from place to place, as the dictates of his work demanded. This was

Johanna's first (semi-permanent) home, and the first of their thirteen children was born there. By all accounts they were very happy.

Bain located a route for the pass which entirely ignored the existing line and struck boldly away up the mountain slope at a steady gradient. In February 1857 he commenced construction with an average of 220 convicts in his team, and by the end of the year the western approach up to the plateau was open for the use of the public. By July of the following year the whole pass, although not complete in all respects, was in general use. As was to become almost his trademark, he constructed a number of dry-stone retaining walls and parapets, and paid particular attention to the various drainage provisions. The Secretary to the Central Road Board, Willem de Smidt, reported to the Board that 'if these drains are merely kept open, the pass will eventually cost the government a mere trifle to keep in repair'.

The pass was officially opened, with fanfares, equestrian processions and speeches, on 17 November 1858, and officially named Grey's Pass in honour of the Governor, Sir George Grey, by the 21-year-old Mrs Thomas Bain. Thereafter all adjourned to the Bains' cottage for 'tiffin'. Judging from the photographs of the cottage, and from what I can remember of seeing it as I passed by when I was road-building in Namaqualand half a century ago, I should imagine that only the most important guests could get into the building – the majority would have had to take their tiffin in the shade of the oak trees in the yard.

The *Government Gazette* of 20 January 1859 contains the official notice of the completion and naming of the pass. The *Cape Argus* carried a less formal report, describing the route as having been 'literally impassable, save at the risk of life or limb ... Now, by the completion of the works which have been carried out under the superintendence of Mr Inspector Thomas Bain, the passage may be performed with the greatest of comfort, for the breakneck kloof has given way to an easy carriage drive.'[7]

This pass was to carry the traffic over the mountains for the next century, until a new pass was completed in 1958, although work began in 1939. The engineers D.K. Bateman, J.M. Hoffman and M.P. Loubser contributed the following reminiscences of the 1939 and 1958 design and construction phases.[8]

'The Piketberg Unit was formed in 1939 to construct the new pass and the road approaches on either side. There was a fair amount of controversy regarding the line on the north side, as Bain's original route ran up the eastern bank of the Olifants River, through Citrusdal and Clanwilliam, whereas the proposed National Road route was up the west bank, bypassing these centres. The National Road Board won.

'Work on the south side started in late 1939, almost simultaneously with the outbreak of World War Two. Result: the closing of the Unit. Work did not recommence until the 1950s. As readers are undoubtedly aware, pass location can be subject to personal preferences, so that staking and restaking of the line kept many young engineers happily employed for some time after the war, undoubtedly benefiting their survey experience considerably.

'The pass, completed in 1958 to a total length of 11.5 kilometres, was surveyed and planned when the 1 in 16 grade was a law of the Medes and Persians, and consequently it had to rise a further 300 feet beyond the neck and could only get down to the banks of the Olifants River several miles north of Citrusdal. When it came to the actual building a colleague pointed out the stupidity of it all. Some sweat was expended by engineers armed with

The old gravel Grey's Pass, circa 1920 (Cape Archives E5317)

aerial photographs and Abney levels, and in the end the present road location was produced, which fits in with the needs of the area ...

'With only a 4-inch base-course on top of a (mainly) sandstone formation, the road section shows remarkable resilience after nearly 35 years of use by ever-increasing heavy vehicle traffic. The latest available information records 2,600 equivalent vehicle units a day, of which 450 are heavy vehicles. Subsidence and signs of wheel tracking are still minimal. Is this perhaps an accolade for the road-builders of the past?'

This new pass was named Piekenierskloof Pass, reverting to the original name, but with spelling updated. It has now been in use for about 45 years and it still carries the traffic travelling the Great North Road to Gordonia, Namaqualand, and, beyond, to Namibia.

✦✦✦✦✦✦✦✦✦✦

Enjoy this lovely pass. Pause at the edges of the top plateau to enjoy the views: to the south over the Swartland, and to the north into the beckoning nest of mountains and valleys fringing the Olifants River, where Jan Dankaert saw a herd of 200 to 300 elephants in 1660. Observe the farming carried out on the plateau itself (and remember the fertile farms on the top of Piketberg mountain, to the south).

A friend of mine maintains that if you want to find water you should look on the top of the mountain, not at its foot – he says the water at the bottom comes from the top, anyway. Be that as it may, Thomas Bain was able to provide a dam on the plateau for the use of travellers, which was undoubtedly appreciated by the draught animals which had just made the climb to the top.

And, sometime, take the road which leads from the top, through a most intriguing lot of mountains to the west of north, finally depositing you at Paleisheuwel. This, incidentally, is the route which Lichtenstein followed in 1803. Why stick to the National Road all the time? I don't!

Thomas Bain (courtesy of UCT Libraries and the Lister family)

CHAPTER THREE

GANTOUW, HOTTENTOTS HOLLAND KLOOF & SIR LOWRY'S PASS

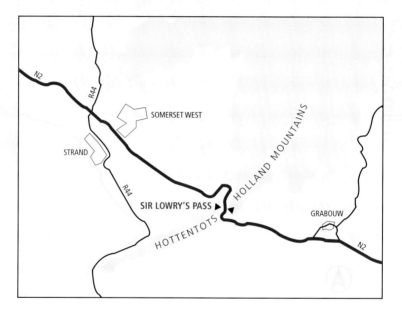

The Hottentots Holland Mountains, running north–south some 50 kilometres east of Cape Town, were the main obstacles to exploration of and, later, expansion along the Overberg, the fertile coastal strip to the east.

The local Khoi herders used the track over the mountains which they called Gantouw (or T'kana Ouwe) to move their cattle as the seasons dictated. This track is just to the north of the present-day National Road, and can still be picked out on the ground.

There is no certainty as to which of the Dutch East India Company's men actually used the track first, but in 1662 Hendrik Lacus, described variously as 'a company surveyor' and as 'a junior merchant, the Fort's second ranking officer', and who had business with Chief Sousoa of the Chainouqua Khoi, is recorded as having ascended the Mountains of Africa via the Gantouw, and then proceeded across the Groenland – the high-lying area past Grabouw – as far as Bot River.[1]

The VOC people translated 'T'kana Ouwe' as Eland's Path or Elandspad. Later, when Somerset West and the Helderberg district became known as Hottentots Holland, the Gantouw was referred to as the Hottentots Holland Kloof, and the mountains as the Hottentots Holland Mountains.

The descriptions left by some early travellers help us to picture this animal track turned stock route turned vehicular mountain pass. Willem van Putten crossed in December 1709. He overnighted in a tent at the foot of the mountain, and in the early hours of the morning 'a strong driving wind from the south east blew our tent down with such force that the tent pegs were dragged out of the ground'. Van Putten goes on: 'half-way up the mountain ... the steepness of the road made it necessary for us [to leave the wagons and] to proceed on horse-back. These willing beasts carried us until we came to even more precipitous heights and we were obliged to dismount in order to avoid accidents and proceeded to scramble to the summit over loose stones.'[2]

It is recorded that a 'Cloevermaker' (the man who maintained the road) was stationed at the pass from the 1740s, and that tolls were raised to finance his maintenance efforts.[3] But when every wagon which descended the pass locked its wheels with *remskoene* for the whole of the descent, his was not an enviable task.

The doughty and well-remembered Lady Anne Barnard came this way with her husband in May 1798. She wrote to Henry Dundas: 'The path was very perpendicular, and the jutting rocks over which the waggon was to be pulled were so large that we were astonished how they were accomplished at all, particularly at one part called 'The Porch' [The Poort]. At length we reached the summit ...'

Captain Robert Percival went over the pass in 1801, and said it was 'sufficient to deter the timid from ever entering the interior of the country ... wild, awful and steep to a very great degree ... the stranger is surprised at finding he has passed in safety.'

Hottentots Holland Kloof in 1776 (Johannes Schumacher)

Wagon tracks on the Gantouw Pass (Cape Archives M927)

William Burchell, in 1811, found

> the road not very steep, but as soon as the traveller enters the hollow way of the Roode Hoogte, the difficulty of the ascent begins. This is a lower hill forming the foot of the mountain, and composed of a hard, barren, reddish, clayey, ferruginous earth, into which the road, towards its summit, is cut down to the depth of, perhaps, twenty feet. After this he has to climb the rock mountain itself, and will not, without some surprise, behold loaded waggons ascending and descending so steep and frightening a road nor will he, without a compassionate feeling for the oxen, witness their toil and labour, carried to the very utmost of their strength; sometimes encouraged by good words, at other times terrified into exertion by the blows of the Shambok, the loud crack of the whip, the smart of the lash, or the whoop and noisy clamour of the boor and his Hottentots ... The danger in which both oxen and waggon are placed while passing the mountains, renders the utmost care and vigilance indispensable. For, should they become restive, and deviate from the proper road, or obstinately refuse to draw, the waggon would be thrown down the precipice, dragging them, and perhaps the driver also, along with it to inevitable destruction.[4]

It was customary for a portion of the toll receipts to be skimmed off for other purposes (as happens to our petrol tax today), but an exception was made in the case of Hottentots Kloof and the Houw Hoek Passes in 1812. Because of the state of the road over these passes, it was 'thought proper to direct that the whole of the receipts of the toll ... shall be in future ... entirely expended upon the road in question, until such time as it is brought into complete repair ... that the safety and convenience of the inhabitants of the interior districts will be attended to, and the facility of communication with Cape Town be materially improved.'[5] But to attempt to maintain an unconstructed mountain pass is a thankless task.

By 1821 more than 4,500 wagons were using Hottentots Holland Kloof each year – and one out of every five was damaged in the pass.[6] Crossing this pass was rather a scramble but, despite the fact that it was extremely steep and dangerous, Hottentots Holland Kloof remained the main route to the east over the mountains for more than 150 years, until Sir Lowry's Pass was built in 1830.

✦✦✦✦✦✦✦✦✦

Sir Galbraith Lowry Cole was Governor of the Cape Colony from September 1828 to August 1833. He immediately appreciated that the development of the Colony was being hamstrung by the lack of decent standard roads into the interior, and especially by the lack of passes over the mountain ranges. By a happy coincidence Charles Michell was appointed Surveyor General, Civil Engineer and Superintendent of Works in the Cape in 1828.

C.C. Michell

Charles Cornwallis Michell, Major (later Colonel) in the Royal Artillery, was a regular army officer. During the Peninsular Campaign against Napoleon's forces he commanded a brigade of Portuguese artillery, and distinguished himself in battle at Badajos and Toulouse. The snippet which I enjoy the most about Michell is the story that he eloped with the 15-year-old daughter of a French officer.

The experienced and willing Michell proved to be just the dedicated and able road engineer needed in South Africa at that time of innovative road- and pass-building. He and the Governor made a splendid team, although unfortunately they were hampered by a shortage of available funds.

In 1828 Sir Lowry told Michell to see what could be done to improve the horrible Hottentots Holland Kloof Pass, as a previous investigation had estimated that it would cost five times as much as it had taken to build Franschhoek Pass. Instead of improving the Kloof road, Michell conceived a plan to build an entirely new pass to the south of the Kloof, cutting across the mountainside at a steady and reasonable grade – and at a cost actually less than that of Franschhoek Pass. Sir Lowry inspected Michell's proposal and was so taken with it that he gave the go-ahead. Preparations for the work were put in hand immediately. Advertisements were placed in the *Cape Gazette* for tenders for food, clothing, blankets and other necessities for the convict labour to be employed on the construction of what became known as Sir Lowry's Pass.

Michell commenced construction with his convicts early in 1829, and the work progressed surprisingly rapidly. For Capetonians it became quite the thing to drive out, or ride out on horseback, with a party to see how the work was getting on. The toll-keeper complained that people were avoiding paying the toll by saying that they were not going over the mountain but just going to have a look, and then not returning. A notice appeared in the *Cape*

View from the summit of Sir Lowry's Pass, with False Bay to the left and the toll house at the top of the pass (Sir Charles D'Oyly)

Gazette: 'To prevent this and likewise to prevent loss of time occasioned to the Miners (who are obliged to leave off work that Persons leading their Horses may pass) it is hereby made known that no one will in future be permitted to visit the Works during working hours.'

The new pass was opened on 6 July 1830, with the triumphal arches and fanfares common at that time. To demonstrate the benefits to be obtained from this engineered construction, Michell and three gentlemen drove up the pass in a wagon drawn by two horses, and then four heavily laden wagons descended the pass without using their *remskoene* – the iron-clad wooden 'shoes' which, attached to chains fastened to the wagon body, were placed under the rear wheels to prevent them turning. (The use of these *remskoene* prevented the wagon rims from being worn away, as they would have been, in one place, had the wheels merely been locked.) These demonstrations were greeted with loud cheers from the assembled populace.[7]

Remskoene were anathema to those entrusted with the maintenance of mountain roads. One comes across a number of tooth-grinding references while paging through old records. Seven months after the pass was opened, Michell reported that the road was 'in a disgraceful state of dilapidation caused by locking of the wheels of wagons', the farmers being 'in the habit of causing the shoe to be put on at the top ... regardless of the very gentle declivity almost throughout the descent ... the shoe having acted all the way as a plough'.

Understandably, Michell wanted the use of *remskoene* on Sir Lowry's Pass, where they were unnecessary, prohibited by law.

In his paper presented to the Royal Geographical Society in 1836, Michell stated:

> Sir Lowry's Pass [was] executed by the orders of that excellent governor, Sir Galbraith Lowry Cole, in 1830, at a very trifling cost, for so I consider the sum of £3000, compared with the benefits which have accrued to the colony generally from this work; and to the neighbouring districts in particular, where double the quantity of grain is now sown, and double the number of waggons, of course, cross the mountain ...
>
> Everyone who has read Barrow, Burchell or other travellers of note, must have been appalled at the very description of the ascent or descent of a waggon by the old Hottentots Holland Kloof; and will feel pleasure in knowing that the same may now be performed at a brisk trot, having become as good a road as any in England.

But notwithstanding all the plaudits Sir Lowry Cole found himself in trouble with the Colonial Office. In his enthusiasm he had given the go-ahead without first obtaining approval from London. Despite being assured that costs could probably be recovered by a toll, the Secretary of State, worried about the finances of the Colony at the time, refused to sanction the scheme and warned Sir Lowry that he might have to cover the cost of the pro-

Michell's gravel Sir Lowry's Pass (Cape Archives AG3327)

ject himself. Incensed, the leading merchants and farmers offered to guarantee any loss he might sustain. Although agreeably pleased by this support, Sir Lowry was not prepared to give up without a struggle. He drafted his now-famous dispatch, which said, *inter alia*:

> The Colony is miserably poor, with a ... population scattered over an immense tract of land, separated from the more civilised parts by mountains over which there are few passes and those of a description that would not be considered passable for a wheel carriage in any other country of the world I believe. Being cut off from a market for their produce there is no stimulus for industry and the inhabitants must ever remain in their present state of poverty and semi barbarism until these passes are made passable.

Finally, rather grudging approval was granted. Sir Lowry later further cleared his yard-arm by writing a diplomatic letter to the Colonial Secretary, Lord Goderich:

> As relates to the expense incurred by me in the new pass over the mountains at Hottentots Holland, I am free to confess as I have stated in my despatch that I deserve censure for not previously asking permission to do so. Having, however, convinced my own mind of its public utility, and ascertained by personal inspection its practicability at an expense I considered inconsiderable, as regards its public advantages, I cannot, however I may wince at the censure I received, regret having taken this step. The Colonists are sensible of its utility and are grateful for the benefit they have already derived from it. But however advantageous a work may be you may rest assured that I shall not readily subject myself to a similar reproof in the future.[8]

So much for encouraging initiative.

Sir Charles Bunbury accompanied Governor Sir George Napier over the pass some ten years after it was opened, and described it as 'an excellent road over this formidable barrier ... the road is narrow but good and its inclination gentle, for a carriage may be drawn down it at full trot in perfect safety'. Bunbury also records that he was told by Michell of a farmer who stated that the new pass saved him a wagon per year.[9]

It is not generally appreciated that the transition from the ox wagon, as the main means of conveyance outside towns, to horse-drawn vehicles was as major a step forward as was the introduction of the motor car in the twentieth century. Trotting horses go twice as fast as oxen. But before Sir Lowry's Pass was opened, the lack of a suitable mountain crossing cut off the Overberg from the more lightly constructed horse-drawn vehicles. Thus, in 1803 General Janssens with twelve oxen to a wagon took 60 hours from Cape Town to Swellendam. In 1837 Sir George Napier did the journey in 30 hours, his wagons each drawn by eight horses.[10]

Michell's line of road was admired by many, including myself, in the following 120 years or so. Its one drawback was that it was initially narrow (understandable in view of the débâcle about costs), so that passing was only possible at selected places. The road was widened on occasion, and especially in 1930, when it was provided with a tarred surface.

But traffic built up, and the pass had to be widened and reconstructed once again, in parts on an improved line. In the early 1950s the easy construction as far as the old turn-off to Sir Lowry's Pass Village (at the hairpin bend) was carried out. Then, from 1956 to 1959 the upper portion of the pass and through to the eastern access to Grabouw was reconstructed at a cost of £291,000.[11]

Dave Scott, who was the Resident Engineer on the contract, contributed the following reminiscences.[12]

The upper section of the pass, rebuilt in 1984 (Hawkins, Hawkins & Osborn)

'In 1956 a tender was accepted for the construction of earthworks and paving of National Road, Route 2, Section 2 (Sir Lowry's Pass and the bypass of Grabouw). The successful tenderer was Simpson Construction. This contract covered a distance of 13 kilometres ...

'Up to that time the road was winding and had a very sharp S-bend at the bridge over the rail tracks on the Cape Town side, and a level crossing at the top of the pass, with a narrow winding road into Grabouw. The new construction straightened the bridge over the tracks, eliminated the level crossing by passing over the railway tunnel and took the now-familiar straight road to bypass Grabouw.

'The most difficult part of the construction was blasting rock to straighten the pass section. The conditions permitted the contractor to close the pass to traffic for two one-hour periods per day, but the contractor elected to use only one of these periods, viz 1 p.m. to 2 p.m. Generally this caused little disruption to traffic after regular users became used to the closure. However, on one occasion when a small amount of TNT was used to 'smooth' the face of the cut, it appeared that the 'lump' was the key holding rock deposits, and hundreds of tonnes of rock slid down, covering the road to the depth of a metre or so and bending the railway line below.

'During that closure the contractor's bulldozer broke its track tensioner at its first try, and the shovel burnt its clutch, so the delay went on until a second bulldozer 'walked' from the Grabouw end. The motorists were not amused. On a lighter note, the driver of an ice-cream truck sold his load to the waiting motorists, while one gentleman regretted that the beer truck was not there as well! It was 7 p.m. before the road was re-opened, allowing an ambulance with an expectant mother to pass.

'Apart from these traffic disruptions, a ganger "supervising" the tree-felling on the bypass failed in his duty and a tree fell across the telephone lines during the budget speech in Parliament, thus cutting off the Eastern Cape and possibly beyond from the broadcast.

'In order to demolish the existing road-over-rail bridge so that the new one could be built, a temporary bridge was constructed with timber felled to clear the new bypass and steel joists hired from the South African Railways for a nominal sum. This reduced the cost of the temporary bridge to £1,000!

'Originally the section between the railway bridge and the summit was to be supported by a retaining wall, but that was changed on site to piers supporting a slab – a bridge over nothing!

'During construction every effort was made to preserve the countryside by replanting on scarred slopes and putting creepers on concrete surfaces.

'Difficulties with traffic were encountered after sub-base and base-course had been laid. Even though a speed limit of 15 m.p.h. was indicated, the new, straight, smooth road was an invitation to some motorists to speed, thus tearing up the surface before it had been paved. The introduction of a double speed hump soon stopped that.

'And a last recollection of events on this interesting job: errors always do occur in tendering, but I think the contractor regretted that he did not do all the price extensions himself, since after he priced $^1/_2$d (i.e. one shilling and twopence) per square yard, the clerk took it to be $^1/_2$d (one halfpenny) or one twenty-eighth of the price for 123,000 square yards!'

'A bridge over nothing': a section near the summit (Hawkins, Hawkins & Osborn)

◆◆◆◆◆◆◆◆◆◆

Again in 1984 the capacity had to be increased to match growing traffic demands. A tunnel was considered, but eventually Hawkins, Hawkins & Osborn designed a sensitive solution to the required widening to four lanes of the extremely difficult upper 1.5-kilometre section, involving viaducts held in place by rock anchors, and other sections which made great demands on the designer's ingenuity. Savage & Lovemore did the construction, greatly constrained by the railway line which parallels the road over a most critical section. The cost of the 1.5-kilometre upper section, and also of rebuilding and widening the existing road over a length of 2.7 kilometres, was R4,413,114.[13]

◆◆◆◆◆◆◆◆◆◆

If you would like to walk up the old Gantouw, it can still be traced on the ground until you top out amongst the eroded rock pillars on the summit.

But you can travel Sir Lowry's Pass in your car at your ease today if you prefer. (Just be prepared for buffeting if the south-easter, the 'Cape Doctor', is blowing.) At the summit pull off into the parking area to the south (plenty of turning room, even with your caravan.) The view is stupendous.

CHAPTER FOUR

Mostertshoek & Michell's Pass

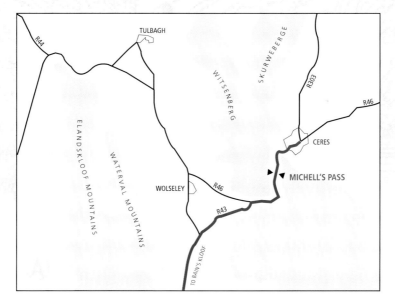

The development of Ceres, 100 kilometres northeast of Cape Town, and of the Warm Bokkeveld and Koue Bokkeveld beyond, was hampered and delayed by the difficulty of crossing the Witzenberg and Skurweberg ranges. It was not until Andrew Bain drove Michell's Pass up the Breede River valley around the southern toe of the Witzenberg, providing adequate access to markets, that the full potential of this beautiful part of the country could begin to be realised.

The development of the route falls logically into five phases.

First came the indomitable pioneers, battling their way across impossible terrain up the Breede River valley to open up isolated farms on the higher ground around Ceres.

The second phase began in 1765 when one Jan Mostert, who farmed the lower reaches of the valley, built at his own expense an eight-mile (13-kilometre) road up the valley, for the use of which he justifiably charged a toll. It was a valued facility for all who passed that way, but it had its drawbacks. Firstly, because it crossed and recrossed the river a number of times, it was impassable after rains. Thunberg had this to say of his 1772 passage of the kloof: 'We had several streams to ford and branches of rivers … These places were the more dangerous to cross as the water not only stood up to the horses' sides, but the bottom was full of large round stones … so that the horses could scarcely get on; frequently the rapidity of the stream was such that they could with difficulty keep the track.'[1]

Secondly, the terrain over which the upper mile and a half made its way was so rugged and precipitous that wagons had to be unloaded, taken to pieces, and carried on the backs of oxen, together with their freight, until the whole lot could be put together and a more normal mode of progress adopted once again. Jose Burman recorded his impressions: 'The old road rises at a horrifying angle, fitted better for mountain goats than oxen. Small wonder that it became the custom to dismantle the wagon and carry it piecemeal up or down this cliff.' Be this as it may, Jan's toll road served travellers for over eighty years.[2]

It is true that there were other, less direct, ways of getting to Ceres. The best of these was via the Witzenberg and Skurweberg Passes, which that adventurous farmer, Field Cornet Jan Pienaar, built out of the Tulbagh Valley in 1780. Although the Witzenberg Pass gave direct access from Roodezand (the Tulbagh Valley) to the Bokkeveld, it was not much of a pass.

In 1822 William Burchell, inspired by the Witzenberg Pass, had this to say: 'The road-makers of this Colony seem to have imagined that, by carrying the road as directly over the mountain as possible, they are following the best, because the shortest, line: but certainly an oblique although longer ascent would, on account not only of easier draught, but also of expedition and even of ultimate saving of expense, be the wisest mode.'[3] Burchell also comments that although the Bokkeveld farmers used the pass, because of its directness, to carry goods to Tulbagh, because of its steepness they did not attempt to ascend it on the return journey, preferring rather to go all the way round via Mostertshoek. It is little wonder that this pass also fell into disuse once Michell's Pass was available as an alternative.

Andrew Geddes Bain, 1864 (courtesy of Errol Lawrence)

✦✦✦✦✦✦✦✦✦✦

The third phase began when Andrew Geddes Bain built Michell's Pass, from 1846 to 1848.

Charles Michell, the Colony's Civil Engineer, inspected and surveyed both the Witzenberg–Skurweberg and the Mostertshoek Passes in 1830. His report, which for some reason was only presented on 23 June 1839, said that the advantages and costs (£3,000) of the two routes were about equal, but that he preferred Mostertshoek. After prodding from the Tulbagh folk, Colonial Secretary John Montagu inspected the Witzenberg–Skurweberg route in September 1845, and offered to build it on a pound-for-pound basis. As the Tulbaghers were unable to come up with their half of the cost, Montagu gave the go-ahead for construction of the Mostertshoek route.[4]

Cape cart on Michell's Pass (Cape Archives J9547)

In 1846 Andrew Geddes Bain applied to Michell for a position as Inspector of Roads just in time to take control of the Mostertshoek construction. Andrew Bain had arrived in the Cape Colony in 1816 at the age of 19. He married two years later, and sired eight daughters and three sons. The family moved to Graaff-Reinet, where Bain worked as saddler, explorer, trader, innkeeper and soldier. During this period he gratuitously supervised the construction of the Oudeberg and Van Ryneveldt Passes to the north of the town, working with the surveyor Charles Stretch. This association enabled him to acquire some theoretical surveying skills from an experienced instructor.

Thereafter he worked for ten years under Major C.J. Selwyn of the Royal Engineers Corps on the construction of military roads on the Cape eastern frontier, including the Ecca Pass between Grahamstown and Fort Beaufort. His basic practical road-building skills were enhanced and refined by the theoretical training he received from Selwyn and his military engineers.[5]

Andrew Bain developed an intense interest and expertise in geology. He produced the first geological map and sections of the Cape Province, and his pioneering monograph, *The Geology of South Africa*, was published in 1851. So meaningful was his work in this field that he has been hailed as 'the father of geology in South Africa'.[6] In 1864, when he visited England shortly before his death, he was honoured and entertained by some of the most eminent scientists of his day.

The Montagu–Michell–Bain triumvirate was to prove the greatest success story in the history of road-building in South Africa up to that time.

Michell prepared the design for the Mostertshoek job, and Bain, with 240 convicts and some specialist assistants – including his son Thomas – at his disposal, managed to complete the construction in two years, a quite amazingly short period. The cost was £22,834.

Bain's report on completion is worth quoting:

> In the place of the old road through Mostert's Hoek, one of the worst and most dangerous in the Colony, a safe and easy pass has now been substituted.
>
> The constant crossing and recrossing of the Breede River is avoided by the new line, which is carried along the right bank of the river nearly parallel with it, till it emerges into the Warm Bokkeveld. For a length of about five and a half miles it is scarped almost entirely out of the solid rock, and is strengthened by a retaining wall, varying from three to forty feet in height.
>
> A massive stone bridge, thirty-six feet in height, and several stone culverts and viaducts, besides a number of drains of considerable extent, have been rendered necessary by the numerous gullies and ravines which intersect the line. With the exception of a hundred yards, which are not yet completed, the whole of this stupendous work has been constructed in little more than two years, by convict labour. That portion of it that has been constructed by blasting has required much labour and exertion, the rock being of rough quartzose sandstone, frequently containing pebbles of pure quartz, in many parts exceedingly hard to drill, and so tenacious as to be very difficult to blast.

The pass was opened on 1 December 1848 by the Governor, Sir Harry Smith, who led an imposing cavalcade comprising John Montagu and other prominent citizens of Cape Town, and most of the farmers of the Bokkeveld and Tulbagh, through the pass. Sir Harry declaimed that it was 'an undertaking which would do honour to a great nation instead of a mere dependency of the British crown'. He then named the pass Michell's Pass in well-deserved honour of Charles Michell, its designer, who had retired a short time previously after being in office in the Colony for twenty years.[7]

It was not only the village of Ceres and the farmers in the Bokkeveld who benefited from this construction. Montagu is on record as having stated to the Central Road Board that the construction through Michell's Pass reduced the time for an ox wagon to travel from Beaufort West to Cape Town from 20 to 12 days – a very worthwhile reduction indeed. It also made express traffic by horse-drawn vehicles easy, a growing need in the Colony during that period as we saw when considering Sir Lowry's Pass in an earlier chapter. Bain's

A section of dry-stone walling on Bain's road (Etienne de Villiers)

Kloof, when built in 1853, saved wagons two more days, and these two passes between them cut transport costs from Ceres to Cape Town to one-fifth of what they had been.[8]

Andrew Bain's 'massive stone bridge' at the foot of the pass, which had been named the Grey Bridge after Sir George Grey, was washed away in 1921 and replaced by a reinforced concrete structure – which became known as the White Bridge! A wider bridge, to meet increased traffic demands, was part of the Phase Five (1988–92) reconstruction.

Andrew Bain's construction thus carried the traffic over the mountain for just short of a century before receiving major attention in 1946.

✦✦✦✦✦✦✦✦✦✦

The fourth phase in the history of this route was the widening and concrete surfacing of the pass by the Ceres Divisional Council, under the guidance and control of the then District Roads Engineer, John Williamson.

The improvement of the road through the pass was long overdue when investigations started in about 1937, the urgency in this regard having been lessened by the provision of alternative transport facilities when the railway reached Ceres in 1912.

Various routes were investigated, but it was finally decided to retain the general location as selected for the existing road a hundred years before. Grades were acceptable throughout, the steepest being three short sections of 1 in 12, and the work in the first stage concentrated on widening the roadway and on improving the curves to meet the needs of the traffic of those days. Most of the old dry-stone retaining walls and some of the impressive masonry

The 1946 concrete surface (Graham Ross)

culverts were incorporated into the new works. Construction commenced in July 1938. Very little equipment was available, and the bulk of the work on this stage was completed by hand labour.

When the widening and realigning had been completed, the Divisional Council had to deploy their unit elsewhere for a year or so before the second stage, the construction of the base-course and the actual surfacing, could be tackled.

Owing to wartime restrictions it was not possible in 1942 to import bitumen for the permanent surfacing of the pass roadway, as had originally been intended. Facing up to this dilemma, the Cape Provincial Roads Department had carried out experiments with thin, unreinforced concrete surfacing, and on the basis of the results obtained on the Provincial Experimental Road it was decided to provide the pass with a 100-millimetre slab on a selected base-course.

Work on the second stage started in August 1943. This was the first time that this type of construction had been carried out on a major project, and much innovative developmental detail had to be sorted out as the work progressed.

The revamped pass was officially opened on 31 March 1946. The costs were just short of £20,000 for the widening and a further £3,370 per mile for the concrete surfacing, giving a total of about £37,800 for the whole project.[9]

The unreinforced slab stood up well to ever-increasing traffic: it was still in use 45 years later. The same basic design was used on a number of jobs throughout the Province: I personally used this design, on a high-class base, on four separate projects.[10]

In 1969 Michell's Pass was closed by landslides caused by the severe earthquake which struck the Tulbagh–Wolseley–Ceres area on 29 September. But when the muck had been shovelled off the surface it was found that the road was still there: it had not slid down the mountainside.

◆◆◆◆◆◆◆◆◆◆

The fifth phase of this historic pass was its reconstruction to improved geometric and materials standards so as to meet the increased traffic requirements of the late twentieth century.

The planning, detail design and construction were carried out departmentally by the Cape Provincial Roads Branch. This was a most tricky project: the work had to be planned and executed in such a manner that traffic was not unduly impeded, especially during the critical fruit-picking season, and a whole number of considerations had to be taken into account, especially when blasting – the railway, major Eskom powerlines and Post Office lines, as well as the fact that the road runs through the Ceres Nature Reserve. Blasting, incidentally, was spread over 34 months and 83,000 kilograms of explosives were used during that time: the 400,000 cubic metres of excavation included 250,000 cubic metres of hard rock.

Restricted space within which to operate with construction equipment, while still accommodating traffic, posed the major obstacle. All *padmakers* hate working 'under traffic': one cannot position or move one's equipment to maximum advantage; non-construction traffic causes delays while one sits fuming unable to get on with things; and, generally, work which

The 1992 road with Bain's construction alongside (Etienne de Villiers)

would otherwise be tackled as a single logical whole has to be split into a varying number of uneconomic little bits. And the situation is exacerbated on a mountain pass where it is generally impossible to construct deviations or make use of existing alternative routes to accommodate the traffic. In the case of Michell's Pass total closure would have introduced a detour involving at least an additional 150 kilometres for the road users of the Ceres district.

Construction commenced during August 1988 and was completed by April 1992. The cost of the project was R42 million.[11] The reconstructed pass was officially commissioned on 20 October 1992. It was nominated for a special award by the South African Institution of Civil Engineering.[12] As you will see when you travel over this pass, the completed construction is very pleasing, from both an engineering and an environmental viewpoint. A section of Bain's original road and a number of examples of his stone walling have been happily preserved for posterity, and I am sure that you will want to climb out, look, wonder and photograph.

In a pleasant ceremony the 150th anniversary of the opening of Andrew Bain's pass was celebrated on 5 December 1998 by a picturesque procession up the pass, speeches and the cutting of a ribbon on a section of Bain's original pass, the unveiling of a plaque commem-

Winter view from Michell's Pass (Keith Young, courtesy of Photo Access)

orating the proclamation of the pass as a National Monument – and an excellent luncheon at the toll house restaurant![13]

The various phases of the development of Michell's Pass demonstrate very clearly how the roads of our country, especially through mountain passes, must precede – or, at the very least, keep up with – and support the growth of South Africa's economy. We must ensure that we do not lose sight of this essential requirement in the years ahead.

CHAPTER FIVE

ATTAQUAS KLOOF & ROBINSON PASSES

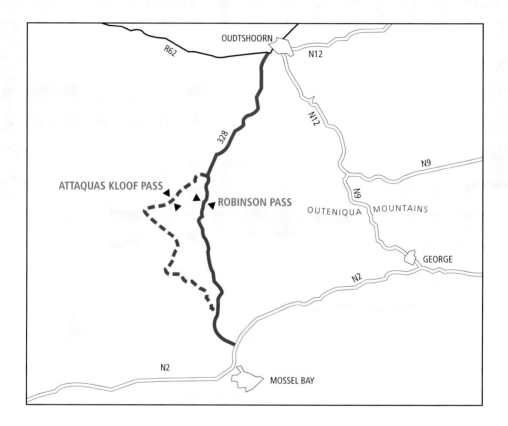

A ttaquas Kloof Pass, about six kilometres west of the present-day Robinson Pass between Mossel Bay and Oudtshoorn, was the earliest recorded route linking the coastal strip with the Little Karoo across the Outeniqua Mountains.

Way back in 1668 Hieronymus Cruse, an official of the Dutch East India Company, led a trading mission to this area. They came by sea (the easiest and quickest way at that time) to Mossel Bay, where he had also to investigate the bay as a possible harbour. His records show that he contacted the Attaqua Khoi 'who lived in a mountain valley', although he himself did not cross over the Outeniquas.

So when Isaac Schrijver came along the coastal strip in 1689, looking for cattle and sheep to buy, and found that the pastoralists had moved into the Klein Karoo in search of better grazing, he remembered Cruse's report. He crossed the mountains through this kloof, widening an elephant track to take his wagons. The crossing took him four days, because of the dense growth through which he had to hack and burn his way. He initially called the pass Lange Cloov but it soon became more logically renamed Attaquas Kloof, after the local people.

Title to the farm Voor Attaquaskloof was granted in 1729, and the pass itself was undoubtedly used on occasions to trek cattle to pasture in the Klein Karoo. But the pass really only came into regular use after the first four loan farms (leen-plase) were granted north of the mountains, in 1756.[1]

Travel eastwards from the Table Bay settlement along the coastal strip was (comparatively) easy as far as George. East of George, however, dense forests prevented vehicular passage. In fact as Charles Michell, the Colony's Superintendent of Works, reported in 1839: 'there is no practical way – not even a footpath, from Plettenberg Bay to the Tzitzikamma country'. In the meantime travellers to the eastern areas had to bypass these forests: they did this by battling inland across the mountains, and then proceeding down the more open Langkloof.

Attaquas Kloof Pass, partly because it was a comparatively 'easy' pass which had been pioneered before the other passes over the Outeniqua and Langeberg, became the established route across the mountains for early travellers. And for over a century the much-travelled Great Wagon Road ran up Attaquas Kloof before swinging into the Langkloof, on its way to Uitenhage, Grahamstown and the eastern frontier.[2]

Attaquas Kloof was superseded to a degree when Cradock Kloof Pass behind George was opened in 1812. But the latter was never really a success, so much so that when on 2 June 1836 Charles Michell reported that 'communications were practically severed between the Western and Eastern provinces, owing to the bad state of Cradock Kloof and Attaqua Kloof' he preferred that 'a small party of convicts, under an intelligent overseer' should carry out minor repairs in Attaquas Kloof, rather than bothering any more with Cradock Kloof – where he considered the only solution to be an entirely new pass.[3]

But this was only a temporary stay of execution for Attaquas Kloof. As far as I can determine, no major engineered improvements were ever made to the road through the pass. Graham Bell-Cross, who made a study of old passes in this vicinity, says: 'it would seem that these types of passes ... were merely formed by driving wagons along the veld ... and entailed little or very limited earthworks. The approaches to all the old passes show a series of wagon roads – probably when one became impassable another was made ... Maintenance in the early 19th century was usually restricted to filling in the odd hole or removing newly-exposed boulders.'[4]

Thus, when the professionally engineered Montagu Pass behind George was opened in 1848, local farmers were virtually the only people who still made use of the grand (if a trifle primitive) old Attaquas Kloof Pass. And even this traffic dropped away when Robinson Pass, only six kilometres to the east, was opened in 1869.

If you would like to walk the old pass you will do well to provide yourself with a large-scale topographic map. Also, arrange for someone to drive around and meet you on the other side: walking the pass twice in one day could be rather a bore.

The Simon van der Stel Foundation erected a commemorative plaque in Attaquas Kloof Pass in 1989, and it was declared a National Monument in 1993.

✦✦✦✦✦✦✦✦✦

Now, De Caepse Wagen-weg (or the Great Wagon Road, as it was known at a later date) had been developed to accommodate the ox wagon of those times, travelling at between four and six kilometres per hour on the level. The traveller needed, besides a road of sorts, regular outspans with shade, water and grazing. These were usually provided 10 to 15 kilometres apart, and were often triangular in shape: they were on the bits of land left over when the circular loan farms were allocated.[5]

In the early 1800s the faster horse-drawn carts and coaches started to replace the ox wagons and saddle horses of yore. There was a call for better roads, and a lesser demand for outspans. This was fine on the flats, but before horses could replace oxen on mountain passes, the pass alignments had to be modified as well as the roadways: what was needed was smaller gradients as well as improved surfaces.[6] This is exemplified by Charles Michell's report after building Sir Lowry's Pass in 1830 that the ascent 'may now be performed at a brisk trot, having become as good a road as any in England'.[7] Attaquas Kloof Pass could not measure up to these standards, so something had to be done to upgrade the Eastern Highway across the Outeniqua Mountains.

ROBINSON PASS

Montagu Pass, opened in 1848, made the crossing of the Outeniqua Mountains a comparatively simple matter, and was compatible with horse-drawn traffic. However, that pass is behind George, and soon the people of Mossel Bay and the Little Karoo began pushing for their own properly constructed pass, which could be used by the new-fangled horse carts and carriages, and which would also reduce the distance from Oudtshoorn to the port of Mossel Bay from the then-existing 135 kilometres to 90 kilometres – quite a consideration, it must be admitted.

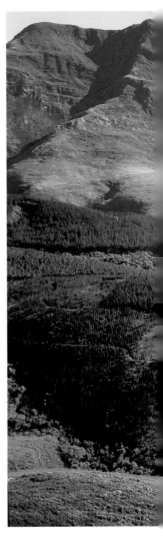

There was a parallel route across the mountains six or so kilometres to the east of Attaquas Kloof, a bridle path not passable for wagon traffic, known as the Ruytersbosch Pad. I do not know when this route was pioneered, but it had certainly been in existence before 1740. It was towards this route that people's thoughts turned when they considered an engineered wagon pass across the mountains.

As a result of various petitions and representations, a parliamentary commission was appointed in 1860 to investigate the possibility of constructing a pass road along the line of the Ruytersbosch bridle (or foot-) path. They estimated that the cost would be £3,682, which parliament proposed to share with the Mossel Bay Divisional Council.

The Divisional Council grasped this straw, and started work, using local labour.

Robinson Pass (Kelvin Saunders)

Unfortunately, what they lacked in experience could not be made up by their enthusiasm, and in 1866 they threw in the towel, having spent only about £700 until then and not having progressed very far.

The colonial government, appealed to once again for help, sent Thomas Bain to sort things out when he had finished constructing Prince Alfred's Pass behind Knysna. He started operations at Mossel Bay in February 1867, just in time to be caught by a very wet winter while he was still building his and his convicts' quarters at the main station.

Bain improved and extended the work so far done by the Divisional Council. He took the road on a new line up the southern slopes of the Outeniquas, working with up to 142 convicts. He reported in June 1869 that 'in addition to the actual construction of the pass, which embraced some very heavy building, the road was extended for five miles beyond its northern base, till joining the old Attaquas Kloof road; thence the old road (an almost impassable track) was substantially repaired and improved by alterations in many places for a distance of three miles to Mr Raubenheimer's farm, from which place there is a tolerably good road to Oudtshoorn.'[8]

The new road was officially opened on 4 June 1869, and was named after the Chief Inspector of Public Works, Murrell Robinson Robinson, who had made the work rather a pet project of his – much to the appreciation of the local gentry. The cost of the pass appears to have been about £30,000, considerably more than the estimate by the parliamentary commission.

The road was reconstructed and surfaced by the Mossel Bay Divisional Council between 1958 and 1962. The route more or less follows Bain's alignment on the south side, but down the northern slopes, where Bain's route had joined the Great Wagon Road in an easterly direction along the Moeras River, the new road shortens the distance to Oudtshoorn by quite a few miles by continuing in a northerly direction. This improvement was carried out as part of the Province's Trunk Road programme.[9]

This is not as spectacular a pass as many another: the curves are generally reasonably sweeping and the grade is not excessive. It is of course surfaced, and it makes a pleasant drive with views of a number of most attractive vistas. It should attract those who do not find themselves particularly enticed by the idea of footslogging over Attaquas Kloof.

CHAPTER SIX

PLATTEKLOOF, TRADOUW & GARCIA'S PASSES

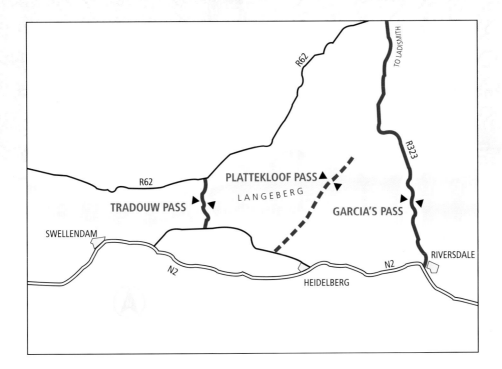

These three passes, which penetrate the Langeberg to link the coastal strip with the Little Karoo in the Swellendam–Riversdale vicinity, are interrelated, and can logically be described together.

PLATTEKLOOF PASS

Plattekloof Pass, behind Heidelberg, was opened up by trekboers, or migrant farmers, about 1740. It appears to have given a comparatively easy passage across the Langeberg, but undoubtedly because of the rugged terrain between its northern exit and the Langkloof, the preferred early travellers' route across the divide remained Attaquas Kloof, further to the east.

However, explorers, naturalists and hunters, as well as the farmers of the vicinity, did use the pass. The first record of travel through the pass is that which Thunberg has left of his trip of 1772, when he was travelling from north to south. Despite this record, when Thunberg wished to go over Plattekloof Pass while returning to the Cape on his second journey, in 1774, he couldn't find the entrance and had to press on to Attaquas Kloof. To even the score as it were, William Paterson, who in 1777 had sent his wagon north up Plattekloof while he himself tried to find and traverse Attaquas Kloof, could not locate that pass and had to return and use Plattekloof.[1] Obviously, not very obvious passes.

The Divisional Council improved this pass in 1841 and again in 1860, when it was renamed Hudson's Pass after the local magistrate – the magistrate was, of course, *ex officio* the chairman of the Divisional Council.[2] Be that as it may, the official cartographer shows the pass on present-day maps as Gysmanshoek Pass – although you will need a sturdy pair of hiking boots to follow the route across the divide.[3]

Plattekloof was rendered redundant for travellers after the construction of the Tradouw and Garcia's Passes in 1873 and 1877 respectively, and fell into disuse save by the local farming community.

TRADOUW PASS

This pass, situated roughly midway between Swellendam and Heidelberg, follows the deep valley of the Buffeljags River across the Langeberg from Suurbraak to Barrydale as the R324.

When the inhabitants of Swellendam wished to get to the Little Karoo on the other side of the Langeberg, or vice versa, they had to travel either via Cogmans Kloof 50 kilometres to

Tradouw Pass (Cape Archives R1857)

the west, or via Plattekloof Pass, 30 kilometres to the east. This of course also applied to Little Karoo farmers who wished to transport their produce to Port Beaufort on the Breede River, which Joseph Barry had opened to shipping in 1841. So we find that in 1858 the folk of Swellendam and district petitioned parliament to have a road pushed through the Tradouw Kloof ('the poort of the women' from *taras*, a woman, and *daos*, a poort), to connect the two communities more closely. They must have had some pull (and of course the Barry family was a power in the land at that time) or perhaps they were just persistent, because in 1867 parliament resolved to construct a pass, using convict labour. The Divisional Council was to erect the necessary buildings, and contribute £1,000 cash.[4]

Thomas Bain was sent out, and surveyed the 13 kilometres of the route that same year.

Tradouw Kloof (Cape Archives R1858)

Chief Overseer C. Hendy started operations with a small party in 1868 by building two convict stations – one at each end of the work – in readiness for the arrival of the main team, and also carrying out the essential prerequisite for all pass construction by opening up a bridle path through the kloof.

Thomas Bain moved onto site in June 1869, and his family settled into the Barry farmhouse 'Lismore', at the foot of the pass. Judging by the writings of Thomas' daughter Georgina Lister, the family was very happy here.

The main body of convicts soon arrived, transferring from the newly completed Robinson Pass. Initially Bain had about 300 convicts, but in subsequent years this number was reduced. Nonetheless, work progressed well, and construction was well advanced when Bain was transferred in February 1873 to build a railway through Tulbagh Kloof/ Nieuwekloof. Overseer Stephens carried the project through to completion.

'F.R.' and his friends traversed the pass in April 1873 before construction was completed.

He described the transit of the unconstructed section thus:

> We induced five of Mr van Colver's men to help us in carrying the cart over that portion of the road which is yet in an unfinished state; and fairly did they earn the sovereign we gave them – for a more 'ticklish' operation I have never witnessed than that of crossing these four or five hundred yards. One wheel had to be unshipped, and then, whilst two men supported that side, the other wheel had to be rolled along the very brink of a sheer precipice three hundred feet deep, the track at some parts not being more than four feet wide. Had any of the men lost their footing our trap must have toppled over and been smashed to atoms. As it was it took just one hour to effect this passage ...
>
> Tradouw Pass or Boschkloof ... may safely claim to be Bain's masterpiece. The gradients are easy, the parapet walls high, solid and continuous, and the road itself is perfect.

The pass was finished not long thereafter, and the opening ceremony was performed by the Governor, Sir Henry Barkly, on 27 October 1873. Although he renamed the pass 'Southey Pass' after a former Colonial Secretary, the locals still called it Tradouw Pass and that is the name by which we know it today.[5]

Patricia Storrar describes Tradouw Pass thus: 'Aesthetically this was a most satisfying pass, full of majestic views alternating with close-up glimpses of the amber-coloured river, dark pools and foaming cascades, with green, rounded hills and shadowy valleys along its length.'[6]

A flood in 1875 washed away 500 metres of the road at the northern end of the pass, and this was rebuilt in 1879 to a higher level. Further flood damage in 1902 and 1906 resulted in a new line over this damaged section, raised to 20 metres above the river. As the old *padmaker* once said: 'There are three things which give you trouble on a road: water, water and water.'

Thomas Bain built a six-metre-span stone bridge over the Gats River as part of the original construction, and this was named Letty's Bridge after one of the Barrys. Unfortunately it was washed away in 1875, only two years after the pass was opened. Bain replaced it with a (higher) teak bridge with a 12-metre span, which still stands proudly today, even though no longer used by traffic: a new, wider concrete bridge was built alongside it in 1979, and named the Andries Uys Bridge.[7]

The pass was reconstructed in 1979, just one hundred years after it was first opened. The 13-kilometre project, which took four years to complete, was most sensitively handled.[8] Pieter Baartman took an 8 mm movie record of the stages of the construction, and John de Kock has recorded his recollections of the difficulties and triumphs of this reconstruction for us.[9]

The decision to reconstruct and surface the pass was taken so as to provide an additional surfaced link between the N2 and the inland, Little Karoo route, and thus complete the planned Trunk Road system in this area.

When planning the reconstruction of this historic pass a number of considerations had to be taken into account, in addition to the normal engineering ones. The pass traverses a particularly scenic fynbos area, which needed to be preserved from damage as much as possible. Every effort was made to retain the attractive old dry-stone retaining walls, and where

An impressive setting for Tradouw Pass (P. Wagner, courtesy of Photo Access)

Tradouw Pass (P. Wagner, courtesy of Photo Access)

new reinforced-concrete walls were necessary these were stone-faced to provide a natural appearance and generally tone in with the old walls. Although the intention was to upgrade the existing road, several deviations had to be introduced to obtain the desired present-day standards: the most important of these required a fill through a side kloof, and this was retained by a gabion wall which merges more readily into the landscape than would a concrete structure. Safety barriers were needed where the road was sited along steep mountain slopes, and here stone-faced concrete walling was provided along the outer road edge instead of the usual metal guardrails.

Another major consideration was the rapid revegetation of the areas exposed by the construction, with plant growth indigenous to that area. Extensive surveys were carried out, but the final line was often determined on site after consideration of the many factors involved. It was finicky work: the shifting of the line by relatively small amounts could both save money and enhance the final appearance and the environment.

The reconstruction of the road from the Suurbraak junction to a point near the Gats River bridge – this is the southern approach to the pass – was carried out by the Divisional Council of Swellendam, and the pass itself was reconstructed by the Provincial Roads unit stationed at Barrydale.

John de Kock concludes his monograph with the statement that 'despite the problems encountered, this road is nicely finished and well worth a visit – the Cape mountain scenery, the fynbos and the road itself will appeal to nature-lovers and the technically minded alike'.

As his write-up demonstrates, it is rare to find a *padmaker* who does not give due weight to environmental considerations when planning and building a road. Traverse the pass, giving yourself time to assimilate all the detail, and you will agree that he and his men did a good job. You will also enjoy the trip.

GARCIA'S PASS

Garcia's Pass, behind Riversdale, crosses the Langeberg through the Goukou (formerly the Kafferkuils) River gorge as the R323 on its way to Ladismith.

Maurice Garcia had come to the Cape from London in 1820, and worked his way up to the illustrious post of Magistrate and Civil Commissioner, first at Richmond and then at Riversdale. He had become fully sensitive to the essential role which communications, and roads in particular, played in developing the economy.[10]

Garcia realised the benefits which the connection of Riversdale with Ladismith and the Little Karoo would have for both districts. He used a few of his convicts to open up a bridle path following the river, and by 1860 this was in general use by horsemen.

Having now demonstrated their willingness to

Maurice Garcia (courtesy of Russell Martin)

help themselves, the locals kept petitioning parliament to construct a wagon road, and in 1872 parliament approved the construction of such a road through the kloof, using convict labour.

Thomas Bain surveyed the route in 1871, and gave an estimate of £3,000 using convict labour. His boss, Chief Inspector Robinson, thought this low, and doubled the figure. A move in the right direction, but not bold enough: the final cost was £29,356 (although it must be admitted that local labour had to be hired from 1875).[11]

Construction started in 1873, with 107 convicts transferred from Tradouw Pass when that job was completed. Garcia's Pass is 18 kilometres long, and Bain also built nine kilometres of approach roads. Construction was not easy – kloofs where water has cut through a mountain range generally have steep side-slopes and little soft material on them. Thus we see a considerable amount of dry-stone walling, as was of course common at that time.

Progress was initially slow, with Bain complaining that the funds allocated were inadequate, but after four years the work was completed. The pass was officially opened on 31 December 1877 by an undoubtedly triumphant Mr Garcia, then in his 77th year. In well-deserved acknowledgement of the contribution by the magistrate and *ex officio* chairman of the Divisional Council in bringing the project to fruition, the link is known as Garcia's Pass.[12]

The next stage in the history of this pass came when the old road was reconstructed from 1958 (with the geometrics being improved in places to meet the requirements of modern traffic) and permanently surfaced. The work was carried out by the Divisional Council with the assistance of District Roads Engineer Louis Terblanche, using Railway Bus Route allocations of between R10,000 and R20,000 per year, which allowed between three and five kilometres to be tackled at a time, depending on the difficulty of the terrain at that point. The upgrading was completed in 1963.[13]

Garcia's Pass is a very attractive road to travel (with or without a caravan). Some of Thomas Bain's original dry-stone walls, up to 15 metres high, may be seen and inspected, as well as sections of the old road, going around noses through which the new road cuts, or going right into side valleys with a resulting turn which would be rather tight for present-day transport vehicles. To my mind, it is a pleasant blending of the old with the new – and it takes me some time to traverse the pass, what with all my stops, lookings, and wanderings off to look at interesting things.

◆◆◆◆◆◆◆◆◆◆

The story of these three passes, Plattekloof, Tradouw and Garcia's, epitomises the continuing struggle which the *padmaker* has to provide the transportation links so essential for the economic development of our country. In this instance the main obstacle to be overcome was the physical barrier posed by the Langeberg, but there are also other, non-physical obstacles which the *padmaker* has to overcome or circumvent in his endeavours to 'make a difference'.

CHAPTER SEVEN

KAAIMANSGAT & DUIWELSKOP PASSES

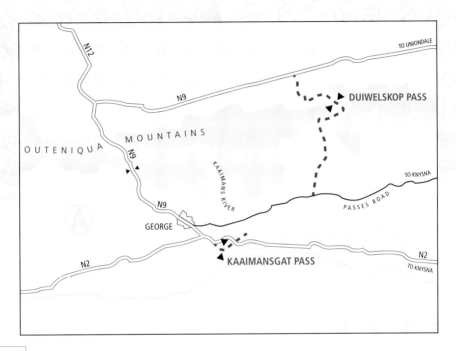

These two passes are interrelated – or at any rate were in their early days – and so both are discussed in this chapter. Both gave access from George to the rich forestry and fishing areas in the Wilderness–Lakes area, and later from George across the Outeniqua Mountains to the Langkloof.

The original crossing of the Kaaimans River was at the 'Gat', over the drift just downstream of where the National Road between George and the Wilderness today crosses the river on a curved bridge. You can turn off the National Road down to the drift. Get out, look around, and imagine taking a wagon down one side and up the other.

The pass has had many names: Quaiman's, Kaymans, Caymmans, Kaaimansgat Drift, Kujman's Kloof – all giving credence to the story that the river was crocodile-infested; or were they just, as seems more likely, that 'large species of amphibian lizard', the iguana or leguan? Wilderness Pass is a more modern name, while Keerom River (or Turnabout River) reflects the feelings of many early travellers on looking down at the daunting prospect presented by the deep gorge and rugged terrain.

Ensign August Frederik Beutler came this way in 1752 on the return leg of his truly epic eight-month expedition to the Qora River and Cradock in the Eastern Cape. He went and

Kaaimansgat in 1816 (C.J. Latrobe)

returned via Attaquas Kloof, but on his way back he received a letter from the Governor when he was at Hagelkraal, near Mossel Bay. He was instructed to search the coast to the east, as there were fears that the French might be trying to colonise the area around the Lakes – these were, as a matter of interest, unfounded. Beutler got as far as Kaaimansgat, took one look at the yawning gulf, outspanned his wagons and proceeded further on horseback. He recorded the Kaaimans as 'Keerom River'.[1]

Although Hagelkraal was at that time the last place of permanent white settlement, Beutler's were undoubtedly not the first wagons to have their passage barred by this river. His is, however, the earliest record which I have found.

The Swedish naturalist Thunberg, on his first journey in 1772, also left his wagon at the Kaaimans, and went on to Knysna on horseback. His description of the area to the east of the river makes one doubt whether it would at any rate have been worth while to take his wagon across. 'The woods we passed through were dense and full of prickly bushes. We could find no other passage through them than the tracks of Hottentots, so that we were obliged almost to creep on all fours, and lead our horses by the bridle.'[2] The forests in this vicinity were extremely dense before the Great Fire of 1869, which raged from Riversdale to Humansdorp, thinned them out. In addition, the rivers in this area cross the line of march, which did not help matters at all. In fact, in one of his reports Andrew Bain, when wishing to demonstrate just how bad a certain road was, said, 'the fearful ruggedness outstrips even that between George and Knysna, and that is saying enough ...'[3]

Governor Joachim van Plettenberg, in 1778, was the first intrepid traveller to take his wagons across Kaaimansgat Pass on his way back from Knysna and Plettenberg Bay, which he had reached by coming south over Duiwelskop. This side trip was accomplished with only three of his wagons, the remaining seven having been sent via the Langkloof and Attaquas Kloof to await the Governor where George stands today.[4] It appears that the area east of Kaaimans had in the interim been opened up to farming – and travel.

From the time of that first crossing, the Kaaimansgat Pass became the accepted route between George and Knysna, at least until Thomas Bain in 1869 opened up the Passes Road with its superior river crossing, four kilometres upstream.

Some of the recorded comments of later travellers bear repeating. Henry Lichtenstein crossed over in 1803:

> This cleft or ravine is one of the narrowest and deepest in the whole Colony ... a steep height is then ascended ... and when arrived at the top ... the monstrous gulph is now directly beneath, and at the depth of a thousand feet below him the torrent roars over its stony bed ... the road now descends and after having crossed the stream, ascends again a height, which as I saw it from this point, I will not say appeared exceedingly steep, it actually appeared perpendicular ... [For] the descent ... the hind wheels of the wagons were locked all the way, at other times all the wheels were locked, and the wagons were partially unloaded, the men dragging after them packages which had been taken out ...
>
> Then begins the most difficult part of the whole passage: to ascend the opposite height ... The task is the most difficult at the very beginning of the ascent, for here the road goes almost as it were in steps ... The greatest difficulty is when the wagon is to be drawn up one of these steps, for in proportion as the leading yokes of oxen get up the steep part upon the level, they no longer share in the draught, so that at length almost the whole of the draught rests upon the hindermost pair. Here then the strength of a number of men must be united to support the wagon behind from rolling back, while the oxen must be compelled to exert their powers to drag it forward ...

Twenty-eight oxen were used to haul each wagon up the final slope.[5]

Even the Revd Christian Latrobe, a respected man of the cloth, who traversed the pass in 1816, had no kind words for it.

> It is not so much its steepness, which renders the passage of Kayman's Gat so dangerous, as the extreme unevenness of the road, if road it may be called, where, as yet, art has not assisted nature, and the traveller must pass over rocks, in steps from one or two feet perpendicular height, the waggons bouncing down, reeling from side to side, and but for the management of the Hottentots accustomed to such service, in continual danger of oversetting. They support the waggon, by thongs fastened to each side, pulling with all their weight, either to the right or left, as otherwise, in several places, the waggons, with all their contents, and the poor beasts staggering before them, would be precipitated into the abyss beneath.[6]

Twenty-six oxen were required to drag Latrobe's wagons up the incline.

> When William Darbyshire's wagons reached the Kaaimans River crossing in 1849, his blood 'froze with horror'. On either side of this deeply gorged river the approaches fell ten inches in every three feet. He stood at the road edge and looked over a precipice of at least 1000 feet. Not a stone or a stump guarded it, not a bush to break a fall.
>
> The drift, 150 yards across, was made of huge boulders which threatened to shake the wagons to pieces. Once across, the travellers were faced with another precipitous climb of more than a mile.[7]

The Swedish botanist Johan Victorin, who left us with many detailed descriptions of terrain, had this to say of his 1854 crossing:

> All at once you came out onto a narrow strip of land with deep ravines on both sides. The road now plunged abruptly downwards, therefore the back wheels had to be locked, and after an extremely steep and stony slope of about one half an English mile we were down on the bottom of the narrow valley at a flowing river called the Kaimans River. It is about 100 alnar [59.4 metres] wide, but only 1 aln deep. The bottom is of rocks as large as buckets and the wagon shook miserably as we went over it. Now the ascent started, which was very difficult ... It is called Kaimansgat. The depth I would estimate at 400 alnar ... Although the straight distance between the tops of the two slopes appeared not to exceed 300 alnar, this short stretch took more than two hours ...[8]

But, let us look at the other side of the coin. Johan Victorin also said, 'He who has not seen this place can hardly get an impression of how beautiful it is ... In a word it was the grandest sight I had ever seen', and Henry Lichtenstein commented, 'The view here is exceptionally good and is perhaps unequalled anywhere else in the whole world.'

It is worth one's time to drive up White's Road behind Wilderness, turn off to the left where the signpost says 'Map of Africa', and look down on Kaaimansgat, look to the right to see the 'map', and look to the left over the sea and shoreline to the east. You will agree, at the very least, that it is a fair prospect.

◆◆◆◆◆◆◆◆◆

Except to upgrade the stony drift itself, there was not much that anyone could do to improve Kaaimansgat Pass until the arrival of the modern techniques and massive road-making machinery of the twentieth century. So it is not surprising to find that a pass was opened over Duiwelskop, probably by the local farmers Jacobus van Beelen and Stephanus Terblanche, in about 1772 to give access to the rich forestry and farming area on the coastal strip east of Kaaimansgat. This was the first pass opened from north to south across the Outeniquas.[9] The fact that people were prepared to travel from George all the way round via the Langkloof and Duiwelskop Pass is proof positive of how great a barrier to wheeled traffic Kaaimansgat was considered to be.

Duiwelskop Pass, also known as De Duivels Kop, Duyvil's Kop, Devil's Kope, Devil's Head and Nannidouw, is about 19 kilometres east of George. It is actually quite an 'easy'

The old Duiwelskop Pass (Kelvin Saunders)

pass, at any rate compared to the other Outeniqua passes. It is one of those passes where the route committed the apparent absurdity of going over the summit of the peak (because the tall wagons could not traverse steep side-slopes without falling over), but other than that and a few metre-high 'steps' near the top the going is easy and the gradients are not excessive – by the standards of that day.

Although he did not traverse this pass but used Attaquas Kloof, Anders Sparrman passed along the Langkloof in 1775 on his trip to the eastern frontier. His comment that the road over Duiwelskop was very bad indicates that there must at any rate have been some sort of road at that time.

Sir John Barrow, who passed this way in 1797, commented:

> From one end to the other of Lange Kloof there is but one passage for wagons over the south chain of mountains and this is seldom made use of, being considered among the most formidable and difficult roads in the Colony. It lies, in fact, over the very summit of one of the points in the chain called the Duivels Kop ...
>
> ... the road was dreadfully steep and stony; and as it approached the summit, where the width of the ridge was not above fifteen paces, the ascent was from stratum to stratum of rock, like a flight of stairs, of which some of the steps were not less than four feet high. Upon these it was necessary to lift the wagons by main strength ... The descent of the Duyvil's Kop was much more gradual than the ascent, and the smooth grassy surface of the northern side was now changed into an extensive shrubbery ...[10]

When Montagu Pass was built the local inhabitants saw what a difference a properly engineered road could make. So we find petitions being submitted for Duiwelskop Pass to be improved. Andrew Bain investigated matters, and in 1856 reported: 'having got so out of repair it has, for several years, fallen entirely into desuetude. It is not the formidable pass I was led to expect from its fearful name and reputation, although it was quite steep enough for all lovers of good roads. But by a little judicious alteration its gradient may be made much superior to any of those intersecting the Knysna plain.'

Nothing was done. The petitions continued. In 1862 Andrew's son Thomas Bain reported: 'to make a new road over the pass would be a matter of very little difficulty as I have seldom inspected a road of such length presenting so few obstacles in construction.' But having between 1844 and 1847 spent £35,800 on building Montagu Pass 19 kilometres to the west, the Central Road Board felt unable to accept Thomas's recommendation to build an engineered pass here to facilitate access to the coastal area east of the Kaaimans River. They did, however, approve that Bain should set out a lower-standard 'boer road' for the Divisional Council to construct, and by 1864 eleven kilometres of the pass had been improved at a cost of about £350.

In the meantime the Central Road Board members had undoubtedly appreciated that, while the frightful road through Kaaimansgat and the very basic pass over Duiwelskop were doing their combined best to serve the area on the coastal plain to the east of the river, the problem really was the lack of a decent road from George across the Kaaimans River and on

The Kaaimans River drift (George Museum)

to Knysna. This would provide the desired access, and would also open up the whole coastal region for expansion.

It is therefore no surprise to find that a Select Committee had already been set up (in 1861) to look into the whole question of a road to the east. The outcome was that construction of this important link started from the George end in March 1867. The new road, which came to be called the Passes Road, crosses the Kaaimans River where the topography is not as dramatic, four kilometres upstream of the old Kaaimansgat drift. Here a crossing was constructed to a much improved standard in 1869. As the 75-kilometre construction progressed slowly eastward, so more and more of the rich coastal plain was opened up for development, with the new road taking over this support function from Kaaimansgat and Duiwelskop Passes.[11]

◆◆◆◆◆◆◆◆◆◆

The Wilderness area, at the western end of the string of lakes, was first developed by Montagu White in the early 1900s, and he built a minor pass, known as White's Road, to give access to the plateau and the Passes Road. In the 1920s, disliking the rather circuitous connection via White's Road, the Passes Road, and over the Silver, Kaaimans and Swart River Passes to George, Owen Grant of the Wilderness repaired the old causeway at Kaaimansgat and cut a narrow roadway for his own use along the face of the slope to connect it to the Wilderness. This is pretty well the line followed by today's National Road.[12]

The National Road at Kaaimans River (Kelvin Saunders)

In 1951, when the Passes Road was unable to meet the growing traffic demands, the main river crossing moved downstream again from Bain's Passes Road crossing point. The new National Road was driven through Kaaimansgat, crossing the river on an attractive curved bridge but unavoidably scarring the approach slopes. The road gave a direct, higher-standard link from George to Wilderness and beyond.

Pieter Baartman and John de Kock have some interesting recollections connected with this construction.

Baartman reminisces about the early staking for formation work, circa 1947: 'Staking the new route to Wilderness progressed well the first few miles out of George; the hard work came when we reached Swart River close to its confluence with the Kaaimans and just above the Kat Falls. On the cliff above the falls we came across wagon wheel ruts in the hard rock as well as a drinking trough cut into the rock surface. We had found the overnight stopping place of early travellers who braved the frightening journey across the mountain via Duiwel's Kop Pass, the first road to penetrate Outeniqualand. Unfortunately construction of our road from the Swart River crossing to link up with the hairpin bend above Kaaimans totally destroyed this historic relic.

'When we reached the opposite side of the Kaaimans we looked for and found signs of wheeltracks climbing the steep hillside, and later believed we had identified the overnight camp site on that side.

'Then, while surveying cross-sections close to the Swart River, I discovered a large cave-like overhang below the staked road-line. I looked for but found no rock paintings though

The Kaaimansgat drift and the National Road (Kelvin Saunders)

there were definite signs of smoke on the overhang ... This must have been a shelter used by the early travellers as well as a Strandloper cave of which several had been identified in this area ... Eventually the blasting of construction all but obliterated my discovery ...'

De Kock reminisces about the later final formation and surfacing period (circa 1951). 'Occasionally one is called upon to do work outside the normal road-building operations. The Kaaimansgat is located at the foot of the low waterfall where the Swart River cascades into the Kaaimans River tidal area. When the original N2 road was built adjacent to the Swart River gorge, a fair quantity of blasted rock, a micaceous schist, landed in the river and was washed into the Kaaimansgat. Most of this work was completed by 1949.

'In 1951, when final formation work was done, property owners at Kaaimans asked that the rock be removed from the Kaaimansgat, as it was a well-known beauty spot but only accessible by small boat. This was agreed to.

'Our cable-operated equipment could not handle the material (hydraulic equipment was for the future) and it was decided to use rafts, made up of timber platforms on 44-gallon drums, with anchor and pull cables fixed to the rock faces. The rafts would be floated in before low tide, loaded by hand and hand-winches, floated out, off-loaded and so repeated until the rising tide stopped work. A considerable volume of rock was removed – gangers and labour gangs often competing to load a raft in the shortest time – and there was no shortage of volunteers. A mass-concrete wall, heavily plummed, was built in the Swart River, above the falls, to prevent further rock being washed down.

'Although the "gat" had been cleared to well below the low-water level, disturbance of the riverbed at times of flood could result in rock movement, and the local people were not completely happy with the possibility, but we could do no more. When this road was upgraded, from 1986 onwards, the widening resulted in more rock landing in the river, but with modern equipment the area was successfully cleared.'

◆◆◆◆◆◆◆◆◆◆

With the need to cater for the ever-increasing traffic this route was upgraded in the late 1980s. At some time in the future, still more capacity will be necessary between George and Knysna. But when a route was being investigated for a dual-carriageway freeway facility in the 1970s, Peter Thomson of Ninham Shand did all the usual things, walked all over the banks on both sides, and even waded and swam through the narrow gorge upstream of the Kaaimansgat drift. He came to the conclusion that the only environmentally acceptable river-crossing point was on the plateau to the north. So it is unlikely that we shall see any but maintenance activities on the present pass through the attractive Kaaimansgat gorge.

As a result of all this road-building activity, Duiwelskop Pass has lost its importance. Today it runs through a forestry area, and is closed to public traffic, but it is possible to hike, mountain-bike or 4x4 through the pass if you first obtain permission.

CHAPTER EIGHT

PAARDEKOP & PRINCE ALFRED'S PASSES

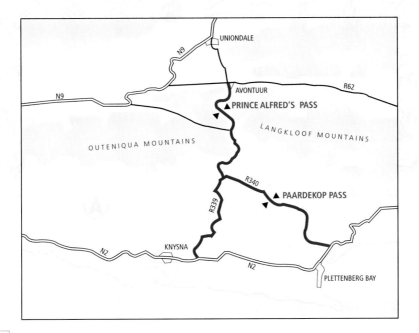

Knysna, with its magnificent lagoon and mountain backdrop, was to all intents and purposes entirely reliant on access by sea when first settled in the eighteenth century. The town's sea connections were, however, rendered a trifle hazardous by the strong tidal flow between the constraining Heads at all times except at slack water. In the days of sail, the fact that these same Heads also blanketed and deflected the wind in this confined passage did not add to the ease of access.

The first vessel to enter the Knysna was the naval brig *Emu* on 11 February 1817. Unfortunately she struck a rock and had to be beached for repairs. But in time ships began to use the harbour safely. A government signal post and pilot was established in 1818, though it was closed again in 1826. The 139-ton brig *Knysna*, built locally in 1831, operated along the coast[1] in what became quite a busy trade: in 1886, for example, 89 vessels traded with Knysna of which two, belonging to the Thesen's Knysna Steamship Service, were permanently on this run. Knysna received official recognition by being gazetted a proclaimed port from 1880 to 1954.[2]

Much timber, as well as other produce, was exported by sea, although in the early days Plettenberg Bay anchorage was generally preferred to Knysna for the loading of timber. But with the spread of railway lines (to Avontuur at the northern end of Prince Alfred's Pass in

1906, the connection from Port Elizabeth to George in 1913, and finally to Knysna in 1928) even this cargo vanished.³ Today sea transport, which originally played such an important part in the economy of this area, is virtually nonexistent.

◆◆◆◆◆◆◆◆◆◆

As for overland links by road, Knysna was for long difficult of access. The dense forests which stretched eastwards from George made approach along the coastal strip from either direction extremely difficult. To the west, between Knysna and George, this forest and the north–south-flowing rivers discouraged wagon traffic. A few rugged and determined individuals did, it is true, manage to drive their wagons between George and Knysna, but there was no decent, engineered road until Thomas Bain and Adam de Smidt completed the 80-odd kilometres of the Passes Road in 1882.

On the other side, a road of sorts was developed between Knysna and Plettenberg Bay, 30 kilometres to the east, but beyond that we can take the word of Charles Michell, who reported in 1839: 'there is no practical way – not even a footpath, from Plettenberg Bay to the Tzitzikamma country.' This was true until Thomas Bain completed the 110 kilometres of what he called his 'Zitzikama road' in 1885.

Prior to this, the Great Wagon Road from Cape Town to the eastern frontier crossed the coastal mountain range, at first via Attaquas Kloof and after 1848 via Montagu Pass, to avoid the impenetrable coastal forest by travelling down the Langkloof. The good folk of Knysna also had to make use of the Langkloof to connect to 'the outside'. Let us see how this route was opened.

PAARDEKOP

There was no road over the mountains directly behind Knysna, but a bridle path (or footpath) is known to have existed up the valley of the Keurbooms River to the east of Plettenberg Bay in 1766, to serve the farms there and to give equestrian access to George. So dire was the need for access to the Langkloof that this was being used by wagons by 1803. As the main destination was George, it is not surprising that this route continued to follow the Keurbooms River when it ran in a westerly direction up the Klein Langkloof north of the Outeniqua Mountains.⁴ For those headed to or from the east, however, a bridle path existed, running north from De Vlugt at the eastern end of Klein Langkloof across the Langkloof Mountains to Avontuur in the Langkloof proper.⁵

The first recorded crossing here is that of Thunberg in November 1772. He was on horseback, his wagons having gone via Attaquas Kloof. In 1803 the wagons of General Janssens (in April) and De Mist and Lichtenstein (in December) crossed from Plettenberg Bay via Paardekop and Klein Langkloof, the first wagon crossings of which records exist.⁶

In 1812 the Commission of Circuit travelled this way and on their return submitted – as required – a report on public roads traversed by them. They had this to say of the Paardekop route: 'This journey [Langkloof to Plettenberg Bay] can be made in two days, and the passage certainly is one of the most difficult and dangerous in the whole Colony; it could

Paardekop Pass in 1816 (C.J. Latrobe)

however be rendered much better at some expense, and the work would be well worth it, it being the common road for the carriage of timber from the Plettenberg's Bay to the Lange Kloof and the adjacent districts and back again for the transport of other produce.'[7]

Now, this route followed by the wagons was not a 'made' road. And like many passes of that era the route had to run more and more nearly at right angles to the contours as the mountain slope steepened and the slope threatened the high wagons with capsizing. Thus we find that because of the steep side-slopes this wagon route went over the very summits of the Groot and Klein Paardekop Mountains.

PRINCE ALFRED'S PASS

In due course it was decided that something must be done to improve the pass, and in 1857 Andrew Geddes Bain carried out a reconnaissance survey with his son Thomas. He was not impressed with the rugged and steep route with which the locals had had to make do for ninety years or so. Andrew maintained that there was no point in attempting to improve the Paardekop route, and that the road should in fact head from Avontuur more or less directly towards Knysna, even though this meant forcing a way through the dense forest along the way.[8] Despite emphasising that this would be 'a work of many engineering difficulties and more costly than anything yet attempted in the whole Colony', with an estimated cost of £15,000, the reports were accepted, and funds were voted.

In 1863 Thomas Bain moved on site and took charge of construction of what was to become known as Prince Alfred's Pass. The work was carried out over a period of four years using up to 250 convicts.

Prince Alfred's Pass (Gunther Komnick)

*Prince Alfred's Pass winds its way along the valley
(Kelvin Saunders)*

The southernmost eight kilometres go through the forest, which for the *padmakers* was very time-consuming. Chief Inspector M.R. Robinson, in his 1862 report, says that the difficult task was aggravated by 'the cutting and felling through a mass of enormous trees, many with trunks 70 feet long and from 30 feet in girth, clearing these huge trunks and logs away and rooting out the corresponding large stumps'. (The Great Fire, which would have lightened their labours on this section, occurred after the pass was completed.)

Once clear of the southern forest area, there is a section of rolling farmland before the true pass section commences, about 50 kilometres from the start. Bain moved himself and his workforce to De Vlugt, where the new road springs away from the easier gradients of the old wagon route up the Klein Langkloof to claw its way up and over the rocky slopes to Avontuur. Bain reported on this section: 'The work is as formidable, and near the waterfalls more formidable than any road yet undertaken in the Colony, and the cost will be very heavy.'

Considerable blasting was required, stone retaining walls – some more than 16 metres high – were the order of the day, and substantial fills were needed to ease the curvature in some of the side valleys. On this section the road crosses the Fuchs (Voogts) River in Reeds Poort (Rietpoort) seven times. Bain used stinkwood beams set in grooves in the solid rock to support his bridges. In 1930 the stinkwood beams, having served for over forty years, were replaced with concrete members.[9]

The work was sufficiently advanced for the post cart to use the new route from September 1866, and May of the next year saw the completion of the pass. Chief Inspector Robinson reported that 'I inspected the whole of the undertaking and I have to express my great admiration of the ability displayed by Mr Thomas Bain in its construction ... the work is, I believe, at least equal to works of this kind in any part of the world.' The cost of the 70 kilometres of new road was £10,632, about £4,000 below the original estimate.

Prince Alfred, the second son of Queen Victoria, visited the Colony in 1867, and in September of that year he was hosted on an elephant hunt in the pass region. Having killed an elephant, he graciously gave permission for the pass to be named after him, and at the official opening on 29 September 1868 the work was named Prince Alfred's Pass. Suitable photographs were taken, and forwarded to His Royal Highness.

◆◆◆◆◆◆◆◆◆◆

The main object of building Prince Alfred's Pass was of course to give Knysna access to the Langkloof, and hence to destinations to the west and to the east.

At the same time as he was working on the main pass, Bain constructed the northerly extension from Avontuur over the Gwarna Range to Uniondale, opening up access also to Graaff-Reinet and the north.[10] Some of the dry-stone retaining walls along this route may still be seen. This mountain pass was replaced by the road built through Uniondale Poort in the 1940s, and this was in turn upgraded in the 1960s.

The link from Prince Alfred's Pass to Plettenberg Bay via the Bitou Bridge, replacing the old Paardekop route, was constructed in 1888 by P.H. Ferreira along a line chosen by Surveyor Charles Lennox Stretch in 1858.[11] As this was an engineered road, it was no longer necessary to go over the Paardekop peaks.

On 29 September 1980 a memorial plaque was unveiled by the Outeniqua Naturalist and Historical Society at the Diep River picnic site in the pass on the 150th anniversary of Thomas Bain's birth. The inscription reads:

> To honour the memory of
> Thomas Charles John Bain
> (1830–1893)
> to whose dedication and
> skill as a road engineer
> the people of South Africa
> owe this fine mountain
> pass and several others
> in the Cape.

Prince Alfred's Pass continues to perform yeoman service. In November 1997 heavy rains caused a section of the dry-stone retaining wall to collapse near De Vlugt. The Roads Department accepted the challenging task of replacing the missing bit of road using the same techniques as Thomas Bain had done. So well was it done that the slightly different colour of the stone-work is today the only clue that the wall is not the original one.

With the exception of a short section at the Knysna end which was reconstructed and given a bitumen surface in the 1940s, and several drifts where concrete causeways have been provided, the pass is largely in its original state. As T.V. Bulpin said: 'There is no way of traversing this pass quickly. By modern standards it is narrow and twisting, but aesthetically, throughout its length it is a complete scenic delight.'[12] He is right.

CHAPTER NINE

HEX RIVER PASS & POORT

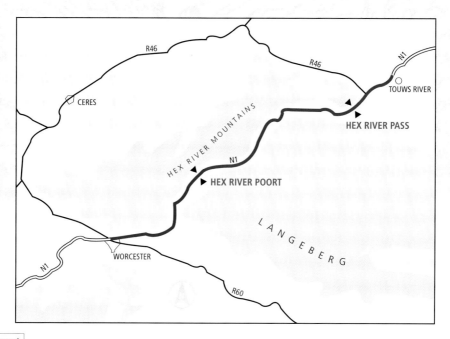

The Hex River valley, 25 kilometres long, lies about 110 kilometres northeast of Cape Town. Access to the western (Worcester) end is via a narrow poort through which the Hex River exits the valley. The eastern (Touws River) end is closed off by mountains, up which the Hex River Pass winds its way.

Once the early explorers and cattle-buyers from Table Bay had crossed the first mountain ranges, via Hottentots Holland Kloof or Roodezand Pass, the Overberg – fringed on the south by the sea – was open to them. But if they wanted to go inland, to the Karoos, they were faced with yet another series of mountains.

There were not many options. Starting in the west, there was the rather frightening Mostertshoek Pass giving access to Ceres, privately built after a fashion in 1765. Next in line, 30 kilometres to the east as the crow flies but over 50 as the ox wagon travels, is the Hex River Poort. This was an obvious way through at least the first range, but with a healthy-looking river which would undoubtedly close off the poort after rains. If our traveller looked further east, the next possibility was Cogman's Kloof through the Langeberg. This was another scramble along a riverbed, in use from the mid-1720s, but not improved at all until the 1850s. And this kloof was 50 kilometres from Hex River Poort, and in addition gave access to the Klein Karoo, and not to the Great Karoo.

So it is easy to see why the Hex River Poort at the western end of the Hex River valley, and the Hex River Pass at the eastern end, despite all their drawbacks, featured in the records of a number of early expeditions, including those of Thunberg and Burchell.

HEX RIVER POORT

In the report on public roads traversed by the Commission of Circuit in 1812 and 1813, which refers to the 'excessively dangerous' Hex River Poort, there is included a statement that a road had been cut out of the mountains by two brothers Jordaan, who 'have rendered the side of the Hex River Poort passable with great trouble and expense'. It was 'very imperfect and wanting the means of repair, and which the Commission found excessively dangerous, while at the same time it is the only way which is capable of being passed on this side'. The Commission recommended the road should be under the direction of the government, as it required widening by a few feet by cutting out the road on the side of the mountain and securing it where needed by a wall on the lower side. The cost of this and subsequent repairs could, the Commission considered, be covered by a toll.[1] Imperfect though the Jordaan road may have been, it appears to have been the first attempt to carry out any but the most basic of improvements to the roadway through the Poort.

The road, the railway and the river jostled through the Poort (Cape Archives AG1041)

The Commission's report concludes with a recommendation which still holds good today: 'a colony like this, where there is such an entire want of navigable rivers, must undoubtedly require everything to be done to promote and facilitate the communication throughout the whole Settlement, by the improvement of roads, establishment of bridges, and the like.' Unfortunately this excellent report seems to have suffered the fate of most reports, and to have been filed with no action being taken. Thus we find the great Civil Engineer, Charles Michell, recording in 1836 that the roadway through the Poort crossed and recrossed the river a great number of times, and that 'in winter this line of road is completely impracticable, as the Hex soon swells into a torrent, rapid, broad, and dangerous'.[2]

Michell, in his paper, illustrates a confusing use of nomenclature. He refers to what I have called Hex River Poort as Hex River Pass. Lichtenstein does the same in his writings. It is obvious that the Poort was, or could be (depending on recent rainfall), the major obstacle to the early traveller. The ascent (or descent, depending on which way the traveller was going) at the eastern end of the valley was just another steep slope to be surmounted in the 'usual way'. So if you find yourself reading elsewhere of the many river crossings in the Hex River Pass, I suggest you consider whether you are actually reading about the Poort and not about what we today call the Hex River Pass. But they are both fully entitled to be referred to as 'passes'.

The first construction by a road authority on the Hex River route took place when the Divisional Council built the road in the 1860s. Jose Burman says that his researches showed that the line had been chosen, both through the poort and up the pass, by Thomas Bain. Unfortunately he does not give his source.

Both the Divisional Council's road and the original route edged through a narrows at the eastern end of the Poort where 'two projecting masses of rock' restricted the available space.[3] A brave decision was taken when reconstruction of the route as a National Road was commenced in 1939, and the line was taken right through this nose. The impressive cutting was known as the Sandhills Cutting (from the name of the railway siding near there and not by any means because the cutting was through sand) and for many years the *padmakers* nibbled away at this mammoth task, which could of course only be accessed from either end.

The road through the poort, the valley and over the pass was reconstructed to improved standards between 1983 and 1986 by Concor Construction to designs by Ninham Shand Consulting Engineers. Mike Judd, who was the Project Engineer, had this to say:

> The Poort itself sat between steep and fairly unstable sandstone and granite cliffs on one side, and the Hex River and the main Cape Town–Johannesburg railway line on the other side. Clearly any attempt to widen the existing roadway, or to construct any bypass for the accommodation of traffic, would be cost-prohibitive. No bypass using existing roads was possible. The need to upgrade the existing road with limited geometric improvements, and to do this under trafficked conditions, became the favoured option fairly early in the project assessment. This, coupled with an assessment of the structural condition of the existing pavement layers, led to the decision that the use of bituminous overlays would be advantageous, and that apart from strengthening the unsurfaced shoulders, no major reconstruction would be done [on this section of the contract].[4]

Any construction under traffic is tricky, but it was particularly so in this case, where the cliff on one side and the drop-off on the other discouraged passing traffic from hugging the outside of the formation as is usual, with the result that the vehicles were even closer to the road-workers than normal.

HEX RIVER PASS

It appears that the earliest wagon travellers climbed the mountain barrier at the eastern end of the valley largely up the bottom of a small kloof,[5] thus getting an easier grade than if they had gone across the contours at near-right angles as was necessary on many early 'passes'.

As mentioned above, the Divisional Council put the first constructed road through in the 1860s. They built through the valley, as was usual at that time, from farm to farm, but at the pass section they were able to get out of the kloof bed and follow an engineered line across the slope. The topography, while not rugged, consists of steep side-slopes. The grade of the road was steeper than is acceptable today, and the old line crosses the present line near the top of the pass.

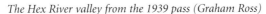

The Hex River valley from the 1939 pass (Graham Ross)

The 1939 construction was to more modern standards, with easier curves and a widened roadway. Through the valley the old, friendly *huisbesoek* line was abandoned, and the road was taken along the southern slopes on a vastly improved alignment. The pass section was also realigned to then-current geometric standards, and the *padmakers* had every right to be proud of their efforts.

However, traffic – especially truck traffic – increased until greater capacity was necessary. Moreover, after forty years of hard wear, the road pavement through the pass where heavy traffic churned slowly up- or down-hill was understandably showing signs of distress.

So in 1979 Ninham Shand started looking at alternatives for the narrow and steep single-carriageway pass section. (The valley section presented no unusual problems – other than trying to avoid dust being raised to settle on the export grapes.) The alignment that was finally chosen and built included a length of split-level dual carriageway with a recommended speed restriction of 70 kilometres per hour in places. The existing road was used by the traffic while the new northbound carriageway was being constructed, after which the two carriageways exchanged roles and the old carriageway was totally reconstructed, several geometric improvements being incorporated to bring things up to date.

This was completed in December 1986, looks extremely attractive and practical, and serves the intended purpose very well indeed. The total cost for the 56 kilometres was R18 million.[6]

All four passes over the years employed the same basic approach to getting traffic up the mountain. Geometric standards have improved from one to the next to cater for the increased traffic demands. Materials design of the constructed road pavement has increased in strength as the ever-growing number and mass of traffic have required. But these are factors which are common to all our major roads and passes if they are to meet the demands for economic growth required to sustain our ever-increasing population.

◆◆◆◆◆◆◆◆◆◆

While this is a book on road passes, I must mention the provision of the rail link through the Poort and valley, and particularly up to the Touws River plateau. This was a most challenging civil engineering work of considerable complexity. The first line was operational in November 1877, with one short tunnel and rather a steep gradient. Subsequently considerable tunnelling and realignment have resulted in the fine rail solution in place today. To fully appreciate it, one must hang out of the window of a rail coach, as I did, or walk the route (which I have not done as yet).

CHAPTER TEN

CRADOCK KLOOF & MONTAGU PASSES

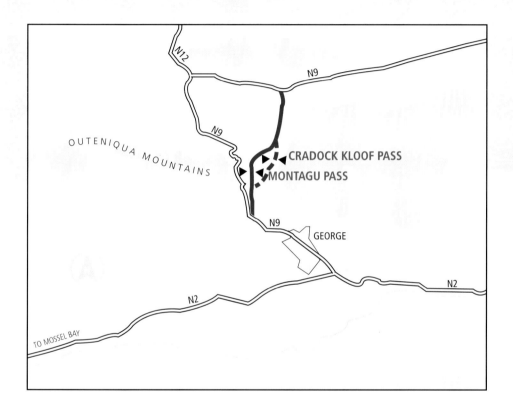

G eorge Town was the first town to be established after the British reassumed control of the Cape Colony in 1806. The town was proclaimed on St George's Day (23 April) 1811 and named after the reigning monarch, George III. It became the seat of a new magistracy, having previously fallen under Swellendam.

Unfortunately for the inhabitants of George, to avoid the impenetrable Tsitsikamma forests, the Great Wagon Road between Cape Town and the Eastern Province crossed the Outeniqua Mountains via Attaquas Kloof behind Mossel Bay to join the more open Langkloof. As a result George remained rather a backwater, unable to take advantage of passing travellers, and the local inhabitants decided that something must be done to provide a pass behind the town.

CRADOCK KLOOF PASS

Adriaanus van Kervel, the first Landdrost of George, created a most praiseworthy precedent when he persuaded the colonial government to accept financial responsibility for the construction by the local authority of a pass over the mountains. The Governor, Sir John Cradock, approved the allocation of 5,000 rix-dollars for this purpose.

The pass was constructed along the line of an existing footpath, apparently in only two months – although there is not total agreement on this point. Frederick Trenck oversaw 40 labourers who worked on the ten-kilometre length in 1812.[1] The road was exceedingly steep, with grades of up to 1 in 4, and the uneven rock strata form a series of 'steps' in places. Allow me to quote Henry Lichtenstein on the subject of these steps: 'The greatest difficulty is when the wagon is to be drawn up one of these steps, for in proportion as the leading yokes of oxen get up the steep part upon the level, they no longer share in the draught, so that at length almost the whole of the draught rests upon the hindermost pair.' Quite often wagoners would in despair take their wagons to pieces and carry these bits and their freight over the pass, assembling them again at the other side.

Records show that the terrible ascent took about three days, with double spans of oxen – if the wagon did not get damaged sufficiently along the way to necessitate delays for repairs. This appears in fact to have been a reasonably common occurrence.

It did not take as long to come down the pass. Andrew Steedman says: 'The view from

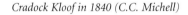

Cradock Kloof in 1840 (C.C. Michell)   *Montagu Pass (George Museum)*

the top of this eminence was altogether as grand as the descent before us was terrific. The wheels of our wagon were locked, and four of the front oxen taken out, in order to render the others more manageable. Notwithstanding this precaution the ponderous vehicle, when once in motion, rushed down the steep and almost perpendicular slope with frightful rapidity. I fully made up my mind that we should not reach the foot of the mountain in safety; in spite of my apprehension, however, we accomplished this perilous descent.'[2]

Charles Michell wrote of Cradock Pass: 'This is an awful obstacle ... You must see the fact to believe that any wagon had ever dared attempt to climb it.'[3] The well-known depiction of the 'State of Cradock's Pass in 1840' was drawn and engraved by Michell, and undoubtedly used by him in his successful attempt to obtain approval for the construction of Montagu Pass.

Generally, then, those who used the pass did not record a high opinion of it. It was delightfully described by one writer as 'fit only for baboons, and for baboons which have the advantage of youth and activity'.[4] Jose Burman probably sums up the general feeling when he says, 'As a road-builder Van Kervel made a good Landdrost'.[5] Be that as it may, Van Kervel had been right about the need for a pass. Between 13 April and 20 October of 1837, 430 wagons, 3,630 men on horseback, 3,210 cattle and 7,130 sheep and goats traversed the horrible pass.[6]

In 1938 energetic youngsters marked the line of the old pass with white-washed stone cairns at the time of the Voortrekker centenary celebrations. If you wish to do so, you may

Montagu Pass in 1853 (Cape Archives AG5855)

walk it, or tackle it on a mountain bike – a chap I met in the mountains told me that it has been used as part of a rally route by his mountain bike club.

MONTAGU PASS

The shortcomings of Cradock Kloof Pass finally persuaded the Central Road Board, under John Montagu, to sanction the construction of Montagu Pass up the adjacent Klip River valley.

Charles Michell, who had recommended the construction in 1839, planned the pass. It was surveyed by Dr W. Stanger, who later became the Surveyor-General of Natal. Construction commenced in 1844. Although some paid labour was used, the main labour was provided by about 250 convicts. H.O. Farrel was the first Superintendent, but he was replaced in 1845 by Henry Fancourt White, a qualified surveyor from Australia, newly appointed a Road Inspector by the Central Road Board. White made his headquarters on the southern foot of the pass. A village sprang up around this hub of activity and was initially called Whitesville, but the name was changed to Blanco at the request of the bashful ex-Australian engineer.

The old toll house (Kelvin Saunders)

The pass is ten kilometres in length, of which about nine kilometres is said to be 'blasted out of solid rock'. Well, it is largely rock side-cut, with fill retained by the dry-stone masonry walls which were such a feature of the passes constructed during this period. It is recorded that £1,753 was spent on gunpowder – dynamite, safer to work with and with about eight times the disruptive power, was only discovered in 1867. Other costs incurred were £4,666 for stores and tools, £8,059 for free labour and £21,322 for convict-related charges, giving a total of £35,799.[7]

The steepest section has a grade of 1 in 6, and is known as 'Regop Trek'. In 1902 Dr Owen Smith and Donald MacIntyre were the first to traverse the pass in a motor car (a Darracq). On their first try they could not make it up Regop Trek. They returned another day and succeeded – with a horse harnessed to the front of the car to provide the needed extra 'horsepower'.[8] But we are getting ahead of ourselves.

The pass was opened to traffic in December 1847. The official opening took place a month later, on 19 January 1848, at the beautiful stone-arch Keur River bridge, which was designed by Charles Michell himself.

The honour of performing the opening ceremony was awarded to Mrs Geesje Bergh, the wife of a former Civil Commissioner of George. 'She drove in her own wagon, accompanied by ten young ladies chastely dressed in white ...' After a few appropriate words to the crowd – which included over 300 mounted men – she announced, 'I christen this splendid road Montagu Pass!' Pneumatic tyres not having been invented yet, she then 'broke a bottle of

Montagu Pass, with the railway pass higher up the slope (Kelvin Saunders)

wine on the parapet of the bridge amid deafening cheers and applause'. Colonial Secretary John Montagu, in responding, said that the pass had been one of the most stupendous and gigantic works ever undertaken by the Central Road Board.[9]

The design and construction of the pass really was an outstanding achievement. Johan Victorin, a traveller of that time, has left us with his thoughts on the matter:

> The road which crosses the mountain range through the so-called Montagues Pass is really a fine piece of work. It follows ravines, often with rock masses overhanging the one side and on the other side abysses several hundred feet deep, against which the still excellent road is protected by a stone wall perhaps one quarter of a Swedish mile long. For about 1 mile it climbs and, after having passed the toll house, again runs downward to the bottom of the ravine, where the Macassar River runs as a murmuring little stream

Montagu Pass near the summit (Graham Ross)

> under a lovely bridge of dressed stone. Then the road climbs again with terrible precipices and sometimes overhanging rocks ... the road had often to be blasted along abysses thousands of feet deep ... The road is not so steep but it climbs all the time without a single flat place.[10]

I found this last an interesting remark, remembering how in my time we provided flat 'breathing' sections on wheel-chair ramps.

The ascent of the new pass could be made within three to four hours with a single team of oxen, compared with the three days (minimum) and double team required for Cradock Kloof.[11] It is no wonder that Montagu Pass was described in an official report as 'fully answering the most sanguine expectations of the public, who hailed that stupendous structure as one of the greatest works ever undertaken in this Colony for the benefit of the community at large'.[12] The pass also has the advantage of being situated in a most attractive area. Emma Murray, the wife of the Revd Andrew Murray, wrote in 1852, 'One forgets everything in the beauty and grandeur of the scene. It was to me exquisite enjoyment.'[13]

The magnificent Montagu Pass marked the commencement of an epoch: it was the first of many passes planned by Montagu and Michell, and the first built under the new system of convict labour necessitated by the emancipation of slaves in 1838. It was declared a National Monument in 1972, and is today the oldest unaltered pass in South Africa.

Montagu Pass carried traffic safely over the mountains behind George for a full century. In fact, it still does, even if the present-day traffic is mainly composed of tourists. However, during this one hundred years the volume of traffic escalated, and the fine Outeniqua Pass was built on the opposite side of the Klip River valley in 1951, and most sensitively upgraded in 1997, to cater for the increasing demand. But that is another story.

CHAPTER ELEVEN

Cats Pass & Franschhoek Pass

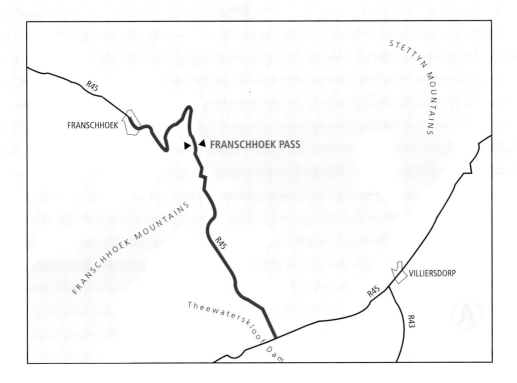

There was no properly engineered road from Cape Town across the 'Mountains of Africa' when Lord Charles Somerset authorised the construction of the Franschhoek Pass in 1823. Hottentots Holland Kloof, 35 kilometres to the south, was a horrible passage, and Roodezand Kloof, 75 kilometres to the north, was only slightly better. But, really, to authorise expenditure on this mountain pass was also rather brave of Lord Charles, as previous ventures by the government – such as the funding of the construction of Cradock Kloof Pass behind George in 1812 – had been far from successful.

Franschhoek Valley lies about 60 kilometres east of Cape Town. It is approached via Stellenbosch and Helshoogte, the latter known in the early days as Banghoek (later Banhoek) Pass because of the wild animals which lurked there and frightened passing travellers. The Franschhoek Valley is closed off at its eastern end by the Franschhoek Mountains, and the pass sneaks across around the northern toe of this range, known as Middagkransberg, between it and the Wemmershoek Mountains.

As with Bain's Kloof, on the western slope it is just a case of deciding on your grade and chiselling your road out of the mountain slope (to simplify the description of the design and construction procedures). The eastern descent is, however, of a different nature. Here the roadway had to be cunningly insinuated into the tortuous contours of a steep-sided kloof, with grand views almost straight down to the kloof bottom, and the design owes as much to art as to science.

THE OLIFANTS PAD

The Franschhoek Valley, where the 176 French Huguenot settlers began their farming and wine-making efforts from 1688, was originally known as Olifants Hoek. At the change of seasons the elephants crossed the mountains and, as was their wont, trod out a defined path for this migration. As on many other mountain passes, the earliest crossings of the mountain by the white settlers followed the 'Olifants Pad', on foot and horseback.

This track was not suitable for wagons without improvements. Governor Simon van der Stel saw it and Helshoogte Pass as being on a desirable route for the wagons bringing much-needed timber from the Riviersonderend Valley to the settlement on the shores of Table Bay, but appreciated that the pass needed improvement to make it suitable for wagon traffic. In his advice to his son, Willem, when the latter succeeded him in the post of Governor in 1699, he included this passage when covering the possible timber routes: 'the one would be by way of Drakenstein, and the ridge or kloof lying behind it, over which, as far as the high mountain range, a road runs, known as the "Olifants pad", wide and broad in some parts, but very narrow at the side of the steep range, and with a horrible abyss below. No wagon could go over it at present, but it is a very short one, and a few slaves might, within three months, so widen it that wagons and oxen could comfortably pass along … It would be the best and shortest road and likewise the most useful. Hence it deserves your attention.'[1]

Unfortunately, for one reason or another the government did nothing to improve the track at that time – or for some time to come, for that matter.

CATS PASS (CATS-PAD)

More than a hundred years passed, and the locals were getting restive because their pleas for a pass to be constructed were meeting with no success. The Stellenbosch landdrost, and his Board of Heemraden, spearheaded the submissions. Finally in 1807 they succeeded in obtaining authority from the Governor, the Earl of Caledon, to build the pass. But, like so many similar petitions, when the project was considered more thoroughly it was found that the cost (8,000 rix-dollars) was beyond their means.

Following various investigations, reports and recommendations in 1813, adequate funds were finally allocated – in 1818. The Board called for tenders for the construction of a road along a line surveyed by W.F. Hertzog. A local farmer, S.J. Cats, was awarded the contract and did his best, completing the construction in 1819. Unfortunately his 'best' resulted in a rough road which was so steep on both sides of the mountain that it could not be traversed by a fully laden wagon. It is in fact recorded that the maximum acceptable load for this pass was eight bags of corn.² A good try by a farmer and Board that had no formal engineering knowledge, resulting in a pioneering route which was in fact probably quite good compared with other mountain routes in the Colony.

Charles Michell, newly appointed as Civil Engineer, Chief Inspector of Works and Surveyor-General, commented in 1828 of the (then disused) Cats Pass: 'It is indeed a subject of astonishment to all who contemplate the still uneffaced rude tracks over which the farmers used to drag their produce to the Cape Town market, that any wagon could ever reach the latter in an entire state.'³ So, unfortunately, this try was not such a success, either.

FRANSCHHOEK PASS

Lord Charles Somerset, Governor of the Cape Colony from 1814 to 1826, was fully alive to the difficulties which the colonists were experiencing in getting their produce to the market in Cape Town. In 1822 he appointed two commissions to investigate the provision of a mountain pass to give access to the Overberg. These commissions were chaired by Major (of Royal Engineers) William Cuthbert Holloway, who headed up the Colonial Engineer's Department at that time. Their reports left no doubt as to the desirability of providing such a facility, either via Hottentots Holland Pass or Franschhoek Pass. Rough estimates put the cost of the Hottentots Holland route at five times that via Franschhoek, so the choice was obvious.⁴

The Governor approved the report and authorised construction of the pass. It is said that his decision to tackle the project was assisted by the fact that there were 150 soldiers of the Royal Africa Corps awaiting shipment to Sierra Leone: '[their] habits were so irregular and cases so desperate' that Holloway suggested they be used to build the pass so that they could 'be prevented from committing violence and depredation on the inhabitants' of Cape Town. His suggestion was accepted.

The Colonial Engineer's Department surveyed the route, aligned it on paper and in the field, and prepared a longitudinal section. 'The acclivity is for the most part not more than 1 foot in 15, and the steepest place, and that only for a few yards in length, does not exceed 1 foot in 7.' The length is a little more than 9 kilometres. In addition to the two companies of soldiers, civilian artificers, at high rates of wages, were also employed on the work.⁵

On the eastern side Holloway built the first stone-arch bridge in the country, with a five-metre span over a kloof named Jan Joubert's Gat. This enabled him to continue along the slope at a reasonable gradient, whereas Cat had taken his road steeply down the kloof to the valley floor below. This bridge, which was incorporated into later constructions, has the distinction of being the oldest bridge in the country still in use, and was declared a National Monument in 1979. It is well worth stopping and climbing down to have a good look at his work.⁶

Estimates of costs varied throughout the construction period. This is really not surprising when it is appreciated that no works of this extent and to this standard had previously been attempted in the Colony. Franschhoek Pass was the first professionally located, designed and constructed stretch of highway in Southern Africa.

The expenditure to May 1823 was £3,830, and the estimated sum 'required for completing the Bridges and Road over this Kloof in the superior manner in which the service has been commenced ... is about £2,400', a total of about £6,230.[7]

Earl Bathurst, Secretary of State for the Colonies, raked Lord Charles Somerset over the coals for approving expenditure in excess of his authority, viz £200, without approval from the Colonial Office in Britain. In reply Lord Charles emphasised the importance of communications and the approaching winter, saying: 'One of the great evils under which this Colony has laboured has been the extreme difficulty of communication with the interior in consequence of the impracticability of the passages across the ridge of mountains which separate the Peninsula from the remote areas.' He also forwarded a detailed report on the work and a statement of costs.

Franschhoek Pass works were finished later in 1825. Major Holloway was able to report that his men had made 'an excellent Road, broad enough for two waggons ... instead of a bane, they have rendered most useful Service, for by their aid ... the formation of a most difficult piece of Road (which, without Military assistance could not have been carried into execution) has been effected.' He reported final costs of the construction of the pass as follows:[8]

Traces of Holloway's 1825 road (left), replaced in the 1930s (right) (NLSA 19555 & 19525)

Construction of 6 miles of road	£5,490–18s– $0\frac{3}{4}$d
Foundations for 2 bridges	£2,260–16s– 1d
Erection of huts	£ 211– 7s– $4\frac{1}{2}$d
Toll house and stable	£ 426–19s– $3\frac{3}{4}$d
Total cost:	£8,390– 0s–10d

But that was not the end of the niggling from far-away Britain. In 1827 Holloway had to reply to a query from the Lords Commissioners of His Majesty's Treasury as to why a supervising subaltern had charged forage for his horse two years earlier. Holloway explained politely that the subaltern needed the horse to supervise the spread-out work parties, and nothing further was heard about the matter.

Franschhoek Pass has been described as 'a splendid mountain road in every respect; the work of that able engineer and gallant soldier, Lieut Colonel Holloway, of the Royal Engineers'.[9] The pass served as the main gateway to the Overberg and the east until Sir Lowry's Pass, more on a direct line, was constructed in 1830. The two routes came together at Boontjies Kraal on the bank of the Swart River about ten kilometres west of Caledon, an early farm frequently mentioned in the journals of travellers at that time and still visible from the N2.

However, traffic from the Franschhoek Valley and Paarl vicinity continued to use Franschhoek Pass. For more than a hundred years Holloway's construction catered to this need with only routine maintenance, until in 1932/33 it was reconstructed with improved geometrics as one of the roadworks tackled during the Depression years. This work was carried out under the supervision of P.A. de Villiers, who later in his career became known countrywide for his National Road location work. Further improvements and the provision of a bitumen surface were carried out in the 1960s.[10]

Today the Franschhoek Pass, besides its contribution to commerce and farming, provides a delightful drive with some quite breathtaking scenery. It is often referred to as one of 'the Four Passes', a well-known tourist day-trip via Helshoogte, Franschhoek, Viljoen's and Sir Lowry's Passes.

If you travel this route, do remember that Franschhoek Pass was the first to be built, that it was the first engineered pass in the Cape (or South Africa, for that matter) and that the Jan Joubert's Gat bridge was the first stone-arch bridge in the country.

CHAPTER TWELVE

THE HOUW HOEK PASSES

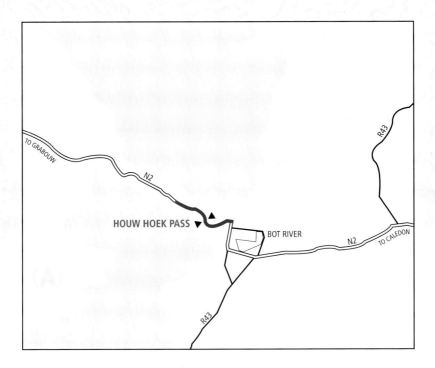

Sir Lowry's Pass was first constructed up the slopes of the Hottentots Holland Mountains behind Hottentots Holland (as Somerset West was called then) in 1830. This was a tremendous improvement on the horrific Hottentots Holland Kloof but it only got traffic up, onto the 'Greenlands' plateau (where Grabouw and all the apple farms are today). If the travellers wanted to go further east, into the aptly named Overberg, they had to get down the eastern edge of the plateau, 25 kilometres from Sir Lowry's Pass. This was the Houw Hoek Pass.

THE FIRST PASSES

It is recorded that Hieronymus Cruse passed this way in 1669, on his way to barter for cattle in the Riviersonderend Valley, and others followed in his footsteps for the same purpose. However, it appears as if they crossed considerably to the north of the present pass. The Swedish mercenary Olaf Bergh in 1682 was the first to cross with wagons (he had two), and for this purpose he needed to follow a different route. He crossed the escarpment roughly in the vicinity of where the present road lies.

William Burchell recorded his impressions when he climbed the mountain here in 1811:

'We came to a short but rugged range of mountains where, by an execrable road, we ascended the rocky pass of Groot Houw Hoek ... This place is much dreaded by those who have to pass it in waggons; though not so steep, it is more difficult than the Hottentots Holland Kloof, and being in the Great Eastern Road, it cannot be avoided ...'

So when it was decided to build an engineered pass up the western edge – Sir Lowry's Pass – it was necessary also to build one down the eastern edge. This Houw Hoek Pass was to become one of Colonel Charles Michell's memorials.

Michell built the first constructed Houw Hoek Pass as part of the same project as Sir Lowry's Pass, but as it was of lesser importance we find that it was finished nine months later than Sir Lowry's Pass, in April 1831. The cost is recorded as having been £600, while £3,000 was spent on Sir Lowry's Pass.[1]

The construction of this pass completed the opening-up of the road over the mountain barrier to the Cape market. As more than 4,500 wagons had battled over the old passes each year,[2] the Overbergers appreciated this very much indeed. The opening was celebrated with 'a trekboer festivity marked by cannon-fire and a gallop to the summit in a horse-drawn vehicle, to confirm that the age of the ox-wagon was drawing to a close'.[3]

One continually finds (if one is looking for them) little reminders that the provision of constructed mountain passes opened up the interior for the lighter, faster, horse-drawn vehicles. This facilitated communications, and made the transition from subsistence to market farming economically more viable, lifting the standard of living in the interior regions as nothing else could or would do. Transportation links were an essential component of the economic development of our country. (In fact – of course – they still are.)

COLE'S PASS

As mentioned earlier, the distance between Sir Lowry's Pass and Houw Hoek Pass is about 25 kilometres. For many years after it was built by Charles Michell in 1831, this section of road across the high plateau was known as Cole's Pass.[4]

In the language of that day, 'pass' had a wider meaning than nowadays. Thus we find in Ordinance 164 of 1805 a reference to 'Passes over the Mountains or Rivers, by which the Produce of the Colony is to be conveyed, either to Cape Town or to any other market'. Cole's Pass was thus the route by which traffic passed over the Elgin plateau. It is included here, firstly, for interest's sake and, secondly, because mention should be made of the difficulties experienced in 'passing' the two rivers of note on the plateau: the Steenbras and the Palmiet.

The narrow Steenbras River, a barely noticed drift in the dry season, has a muddy bottom which softened up and caused quite considerable delays after rains. Tree trunks and boulders dumped in the drift to provide a firm footing proved to be only temporary palliatives. The construction of side drains in 1853 reduced the flooding over the drift itself, thus lessening the build-up of mud over the hard surfaces.[5] A culvert was constructed at the crossing about 1920, much to the relief of those daring motorists who had previously crossed the drift with the water deep enough to cover their floorboards.[6] In the 1930s a proper culvert was provided at the same time as the route was 'tarred', and nowadays of course we cross on the National Road at the headwaters of the raised Steenbras Dam.

The second river was the Palmiet, 'named by the Hottentots Koutima or Snake River' – look at its sinuous alignment for an explanation.[7] There were many occasions when travellers were forced to wait until the flood waters subsided. Sir John Barrow recorded that in 1798 the river was impassable for four months in the year. The following year an impatient Assistant Surgeon Patrick of the 8th Light Dragoons was swept away while attempting to cross on his horse, and drowned.

A pontoon was in operation in 1801. The English visitor Robert Semple says that the travellers' horses swam alongside the pont, while the would-be riders sat in the boat, holding their reins and keeping the saddles dry.[8] Finally, in 1808 a bridge was built, which was the first bridge in the country outside a town. The Revd Christian Latrobe leaves us with a good description of the structure as he saw it in 1816:

> On the 14th of June we passed over the bridge, which is of wood and the only bridge in all South Africa. It rests upon stone piers, though sufficiently strong to resist the force of the stream which sometimes in the rainy season becomes extremely rapid and fierce, but was now nearly hidden among palmite-plants and low bushes growing in its bed. To prevent the woodwork being washed off the piers in great floods, the timbers are fastened to them by strong chains, on the side of the water's descent. The bridge is furnished with railings, and on the floor planks the thick spungy stalks of the palmite-plant are laid in abundance, partly to provide an easy passage for the bullocks' feet, and partly to deaden the sound of the wood by which they are apt to be frightened.

This was the Oudebrug. In 1852 a new bridge was built two kilometres upstream, on the line of the main street of Grabouw town. The National Road now crosses on a massive bridge, built in 1958, a short distance upstream of the Oudebrug site.[9] The vexatious Palmiet River has finally been tamed. And Cole's Pass (if anyone remembers the name) is passable.

HOUW HOEK PASS

It will be remembered that Sir Lowry Cole was criticised for incurring expenditure on pass-building without approval from the Colonial Office in London. It appears probable that as a result Michell was instructed to make as cheap a job as he could of Houw Hoek. At any rate, his construction did not stand up very well, for we find Andrew Bain reconstructing the pass 15 years later, in 1846, at the same time as he was building Michell's Pass. He spent £4,211 on the job – considerably more than Michell's £600.

Bain improved the alignment in places, and constructed a number of dry-stone retaining walls to assist in this endeavour. He also undoubtedly introduced engineered drainage, and provided a selected gravel riding surface, as with his other passes. Today, as you descend the pass you first come to an easy left-hand hairpin bend. Then, just as you are feeling it is safe to build up a bit more speed, you come to a tighter right-hander. Very shortly after this a road takes off downhill, to your left, going to Bot River village: this was more or less Bain's line.

A report from the Central Road Board, in the *Government Gazette* of February 1847, states: 'The new pass of the Houw Hoek was opened to the public in January last, to the great joy of every person who had experienced the severe trials of the old.' People, like Charles Dickens's Oliver Twist, are always wanting more. The 'old' pass had been greeted with 'great joy' – and cannon-fire, etc. – only 16 years previously.

Bain's pass was in use for fifty years or so, until shortly after the Anglo-Boer War.

One of the danger spots on the previous pass had been the climb up to and the steep descent from the mountain top above Bot River town. So when the railway was pushed through the kloof, it showed where the road could also go. Sections were rather tortuous, earning the name Poespas, or Higgledy Piggledy Pass, but crossing over the summit of the mountain was now avoided.[10]

The road was well built, as would be expected of work carried out at that time. But its geometrics were not suited to the demands of car and truck traffic.

The 1930s road (right) (NLSA) replaced the old 'railway pass' (below) (Cape Archives R1068 & R1070)

It must be remembered that rural motor traffic was still almost unknown in the early days of the twentieth century: for example, it was only in October 1902 that the intrepid T. Silver piloted the first motor car to travel all the way from Cape Town to Port Elizabeth.[11]

The growth of motor traffic necessitated the reconstruction or improvement, to more suitable and safer geometric standards, of many passes on the main routes. Houw Hoek Pass could not be meaningfully improved through the railway pass, and in the 1930s it was re-routed back over the mountain summit. Obviously the alignment was improved, and advantage was taken of the more powerful road-making machinery available then to overcome physical obstacles such as dongas, valleys, spurs and noses. A very fine road resulted, a great improvement on the railway pass.

But it has frequently been remarked that traffic has a habit of increasing over the years, and in 1976 Houw Hoek Pass was again reconstructed, with four lanes on dual carriageways, and with the improvements to the geometrics which this approach enabled.

Both design and construction may be admired while traversing the pass. And of course the views over the lower adjacent Overberg are quite breathtaking. There is a pull-off at the second hairpin bend where you can relax and enjoy the pleasing prospect.

CHAPTER THIRTEEN

Bain's Kloof Pass

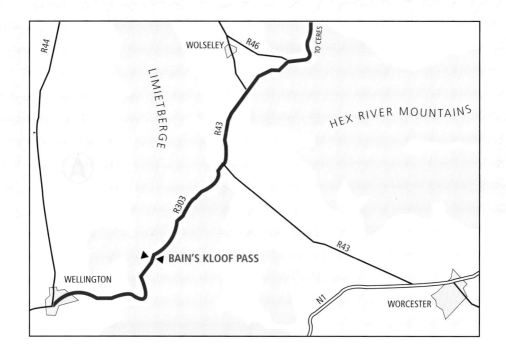

Bain's Kloof Pass, which crosses the Limietberge to the east of Wellington on the road to Ceres and the north, was opened in 1853. It is a work of considerable engineering complexity, which is generally regarded as being the magnum opus of Andrew Geddes Bain, the famous road-builder and geologist.

In 1846 Bain was appointed as an Inspector of Roads in the Western Cape, working under Charles Michell, on the impressive programme of road-building conceived and implemented by John Montagu.

One of Montagu's ambitions was to open up a direct line of communication with the interior. So, while engaged on his first major project, the construction of Michell's Pass, Bain gave thought to the possibility of building a road cutting straight across the mountains to Wellington instead of all the way round the mountains via Tulbagh Kloof. Riding back to Michell's Pass along the Breede River valley with Montagu from an inspection of Houw Hoek Pass (where Bain was also supervising various improvements), he pointed out to the Colonial Secretary a promising-looking kloof where the Witte River debouched between the Limietberge and the Slanghoekberge, which he considered led in the direction of Wellington. Montagu is alleged to have cried out, in great excitement, 'Bain! That's just the line!'[1]

Montagu Rocks (Cape Archives)

After that there was nothing Bain could do but ride to Wellington at the earliest available weekend, and make enquiries of the local inhabitants about a way through the mountains. True, there was a bridle path crossing the Limietberge about eight kilometres north of where he wanted to go, which went down the eastern slope in Bastiaanskloof, joining the Witte River valley about midway between Tweede Tol and the Breede River Bridge, but a look at a contour map will show why both alignment and topography eliminated this option.

Bain found no support for his theory that there must be a way across and down the Witte River valley until he met Johannes Retief, who had been a distance up the Witte River kloof from the Breede River end and also had gone a short way down this valley from the Wellington side.

On this Saturday afternoon Bain examined the western side of the mountain, and found no difficulty in tracing two quite easy lines for a road up to the neck. Suitable financial arrangements having been concluded, he engaged Retief as guide, and enlisted the company of two sons of Mr Daniel Malan (a member of the Stellenbosch Divisional Council Board) and Mr Septimus du Toit. Leaving at 4 o'clock on Sunday morning, they rode on horses provided by Field Cornet Rousseau up a cattle track to the neck. Their horses were then sent round via Bastiaanskloof, there hopefully to await their arrival down the kloof.

After an obligatory period spent admiring the magnificent view to the west, from whence they had come, they commenced their scramble eastwards down the kloof. Andrew Bain's description bears quoting:

In this frightful terra incognita everything is repulsive and savagely grand, and may well account for the ignorance that prevails on both sides of the mountains respecting that locality. Vast piles of funereal-looking rocks everywhere disturb your progress and the black disjointed krantzes, that every now and then protrude their unearthly shapes to the very brink of the precipitous banks of the foaming torrent [the Witte River], seem for ever to have set at defiance the approach of man. For the first three miles we had nothing but crossing and recrossing the river and climbing up the mural banks at the risks of our necks and, so gloomy was this place, there was a perfect absence of animal life ...

However, with all those apparent disadvantages, I could distinctly trace a line for a road along the left bank of the stream ... which is perfectly practicable but will be very expensive ... We had to skip from one stone to another and force our way through thick brakes, wade the river times without number, climb precipitous krantzes, tumble down again among loose stones and, on the whole, such a fatiguing exploration I never had before, for when I returned hither about sunset, having got my horses at Bastian's Kloof, I was completely exhausted and done up and, indeed, I have not yet recovered ...

On the whole, I look upon the Kloof as a grand discovery. It will facilitate your idea of a direct communication with the interior and bring the main road where it ought to be, through the most populous part of the Colony.[2]

Dacre's Pulpit, 1878 (upper right)
Montagu Rocks (lower right)
A general view (lower left)
(Cape Archives AG1691, AG17028 AG1692)

An enthusiastic Montagu replied, 'Your letter quite delighted me. Bain's Poort will be our next job, so get Mosterts Hoek out of hand as soon as possible ...' But for all that, the Assistant Surveyor-General, Charles Bell, carried out a confirmatory inspection before work commenced.

What became known as Michell's Pass through Mostertshoek was formally opened in December 1848, and when Bain had finished tidying up there, and when the accommodation, convict barracks and kitchens, built by civilian artisans, were ready, he moved the remainder of his team over to Bain's Kloof. To these he added an office, storerooms, magazines, a hospital, a church cum school and a recreation area, stables and blacksmiths' and carpenters' workrooms. The main station was at the neck initially (it was moved to Tweede Tol on the eastern side in 1851 when the emphasis changed), with subsidiary stations along the line. As Mackenzie comments, 'It was, no doubt, an impressive camp, all neatly whitewashed.'

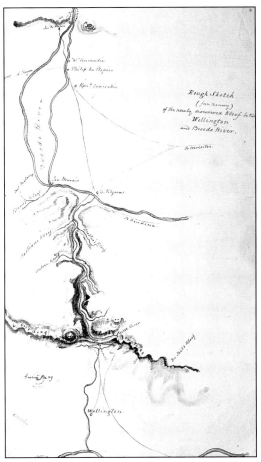

Andrew Bain's sketch of Bain's Kloof (Cape Archives)

This project was by far the most venturesome to be tackled in the nineteenth century, and was the one which had the greatest number of convicts engaged on it. They generally numbered between 300 and 350, on occasions touching the 450 mark. Owing to completion of and remission of sentences, more than a thousand individual convicts were employed on the pass during construction. Work was started on 16 February 1849, and the pass was formally opened on 14 September 1853, four and a half years later.

Initially Bain worked on the easier western approach, as was logical, to secure his supply lines. Little blasting was needed on this side. The road was trafficable to the summit by the end of the year, although only considered to be 'completed' two years later. There were two timber bridges on stone piers, and four smaller stone culverts along this section, and the enthusiastic *padmaker*-environmentalist lined the road with over 300 oak trees.

Halfway up this section Bain constructed the first road tunnel in South Africa, cutting across a nose to shorten the route. According to Bain's sketch it was 400 feet long, 16 feet high

Dacre's Pulpit, 1965 (Graham Ross)

and 12 feet wide (122 m x 5 m x 3.7 m), built on a three per cent grade.[3] Unfortunately the *in situ* material, of which Bain said 'the tunnel cuts like cheese', proved unstable and liable to rock falls. After the heavy winter rains of 1850 caused major falls to block it, the tunnel was abandoned, as also was the proposal for a second tunnel near the crest. The two portals of the failed tunnel were signposted by Divisional Council engineer M. Austin in 1988, and may still be inspected.[4]

The eastern slope was an entirely different proposition, at any rate in its upper reaches. To gain access Bain first had, as is usual with a new alignment, to open a bridle path along the route. Work on the approach road, and on the hard rock-cuts and dry-stone retaining walls on the eastern slope, started in 1850. The same heavy winter rains which spelled the demise of the tunnel on the other slope caused damage to the east, which it was estimated set progress back three months.

Bain described the changed nature of the work thus: 'The nature of the work here is quite different from the other side of the mountain, the line passing through masses of fixed and detached quartzose rock which seem to set at defiance the engineer's skill to construct anything like a well graduated road through it: for no sooner is one obstacle removed, in the

Clinging to the slope above the Witte River (Graham Ross)

shape of an enormous block of rock, of scores of tons in weight, than others appear in rapid succession, such as transverse rocks etc. of which there appears to be no end; but the powerful agency of gunpowder is slowly making them disappear.'[5]

All drilling had of course to be done with a hand drill and sledge – the junior man holding the drill steel – and the blacksmiths were kept busy resharpening the drills. Mackenzie reports that the remnants of holes which he measured had a diameter of 40 millimetres.[6] Wedges were used to split boulders – it was only at a later date (when work was on the go in Meiringspoort) that Bain developed the method of building a fire around the rock and then dousing it with cold water when the rock was hot, to cause the rock to split. Large rocks were moved with muscle power, steel bars and, sometimes, rollers. Bain recorded that he used a short section of rail at what is known as Montagu Rocks to remove the broken rock from the box cut where the excavated rock couldn't be pushed over the edge.

Virtually the whole of the first ten kilometres from the top had to be blasted out of solid rock. It is difficult to judge the depths of side-cut necessitated by the road width as Bain on many occasions blasted additional rock loose from above the road to provide material for retaining-wall construction, but 20-metre cut faces would appear not to be uncommon. Some quite considerable (for that time) box cuts were made through solid rock noses.

Again, long lengths had to be supported on the downhill side by dry-stone retaining walls, some up to 20 metres high. These are still holding the road up today, 150 years later. Bain described one such section as being of 'the most appalling and difficult kind; the lofty retaining walls being built on the very edge of a precipitous cliff 300 feet high whilst the upper half of the road is blasted out of and stolen as it were from the frowning krantzes above'.

Once clear of the rocky kloof the nature of the work changed again, and picks and shovels replaced drill steel and sledges. Two major bridges, later named Pilkington and Borcherds Bridges, had to be built on this section of the work.

It was also of course necessary to build a road linking Michell's Pass to the Kloof – which included the Darling Bridge over the Breede River – and another road on the other side linking the Kloof to Wellington – where a bridge was built over the Berg River to carry traffic beyond the town.

Bain's Kloof Pass was officially opened by the Chairman of the Central Road Board, Petrus Borcherds, on 14 September 1853 with due pomp and ceremony, the drinking of toasts to all and sundry, and many speeches. It was on this occasion that Andrew Bain, in replying to a speech of praise when it was formally announced that the pass was to bear his name, made his famous statement, 'I would rather make another road than another speech ... being but a common highwayman more accustomed to blasting and blazing.'

The 30-kilometre pass had taken four and a half years to complete and cost about £50,000. It is in use, virtually unaltered, a century and a half later.[7] In 1934 the pass was tarred, with the opportunity being taken to effect minor improvements to the alignment on the Wellington side. The cost was £40,000.

Bronze plaques at the neck commemorate the centennial celebration of 1953 and the declaration of the pass as a National Monument in 1980.

◆◆◆◆◆◆◆◆◆◆

It is perhaps not generally appreciated today that the main coaching and transport road to the north passed along this route and via Sutherland, Fraserburg and Victoria West in the nineteenth and early twentieth centuries.[8] The opening of the diamond fields in Kimberley in 1870 consolidated its claim to be the most important route from Cape Town to the interior.

In 1925 the Holmswood Commission was appointed to report on the 'present road and bridge position in the Union and to make concrete recommendations regarding a future road policy'. This excellent commission proposed a system of carefully chosen National Roads to be constructed to a high standard over ten years. One of their recommendations was that the route from Cape Town to the north should continue to run via Bain's Kloof, Ceres, Karoo Poort, Sutherland and Carnarvon to Johannesburg.[9]

It was only with the promulgation of the first National Road five-year-plan in 1936 that, after considering a myriad of submissions, and hearing what many deputations had to say for and against both routes, and after prolonged and oft-times bitter discussions, a reluctant decision was taken to give preference to the route via Worcester and Beaufort West. This decision was clinched when Du Toit's Kloof Pass, built to standards more suited to modern motor traffic than Bain's Kloof Pass, was opened in 1949, and upgraded by the construction of the Huguenot Tunnel in 1988.

But that is another story altogether.

CHAPTER FOURTEEN

MEIRINGSPOORT

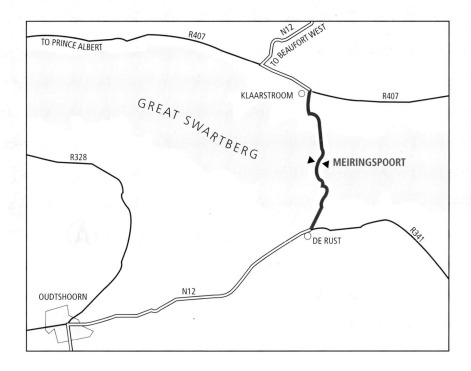

The Great Swartberg range has always been a barrier to communication between the Little and Great Karoos. Until the first road was pushed through Meiringspoort, 35 kilometres east of Oudtshoorn, it could in fact have been described as being almost insurmountable.

The first recorded transit of the poort was by a farmer of De Rust, Petrus Johannes Meiring, in 1800. In due course he and Gerome Marincowitz of Vrolikheid, near Klaarstroom at the other (northern) end of the poort, opened up a bridle path along what was then known as De Groote Stroom.[1]

In 1854, as a result of petitions received by the government, Sir John Molteno, then MP for Beaufort West and destined to become Prime Minister, Andrew Geddes Bain, his son Thomas Bain, and Charles Pritchard, a Beaufort West lawyer, travelled from Beaufort West by horse – a journey which took a few days – and carefully examined the entire route.[2]

Another route through the Swartberg – or, strictly speaking, between the Swartberg and Slypsteenberg – was possible through Toorwater Poort, 50 kilometres east of Meiringspoort. Thomas Bain considered this route infinitely superior to that through Meiringspoort from an engineering viewpoint: it was only five kilometres long and only three river crossings

Cape cart at a drift, circa 1885 (Cape Archives AG1063)

would have been necessary. He estimated that a permanent road could be constructed for only £5,500.[3] However, Toorwater Poort is, as said, about 50 kilometres east of Meiringspoort, which would have meant an extra one hundred and more kilometres' travel from the Cape or Mossel Bay, or four to five days' travel by ox wagon, so it is small wonder that Toorwater Poort was left for the railway. In due course the rail line between Oudtshoorn and Klipplaat junction was built through the poort.

As a result of the Molteno report, a Select Committee was appointed in 1856 to investigate the proposal to build a road through Meiringspoort. They had to choose between a 'boer' road, which would be subject to frequent wash-aways, for £3,000 to £4,000, and a properly engineered road for £50,000. They allocated £5,000!

The route was surveyed at various stages by a Mr Woodifield, Adam de Smidt, Andrew Bain and later Thomas Bain,[4] and in August 1856 work commenced with 93 hired labourers under the supervision on Thomas J. Melville, Sub-Inspector of Roads in the George district. After complaints were lodged Adam de Smidt took over, under the general control of Thomas Bain, his brother-in-law.

It was here that Andrew Bain developed the rock-splitting technique which has been used by generations of *padmakers* since then. He reported: 'After the bushes [which obstructed the road] had been cut down I had them heaped up on the large detached rocks which every-

The 1858 road above the 1971 road (Malcolm Watters)

where impeded our progress, and set fire to them, the effect was quite magical; especially after throwing buckets of water upon [the rocks] they become quite brittle and crack like glass. This plan alone will save an immense sum in blasting; very little of which, by adopting this cheap manner of removing the rocks, will be required.' Judging from my (admittedly meagre) experience of this approach, I would hesitantly suggest that Andrew's last sentence is a trifle optimistic.

The 16 kilometres was constructed in 223 work days, crossing the river 21 times, at a cost of £5,018. The pass was officially opened on 3 March 1858, Colonel A.B. Armstrong, the Civil Commissioner of Oudtshoorn, breaking a bottle of Bass's Best Ale and naming it (at the suggestion of Andrew Bain) after Petrus Meiring, who had campaigned so tirelessly for the road to be constructed. A procession of 300 horsemen and 50 carts carrying smartly dressed ladies then traversed the length of the poort, and twelve wagons, heavily laden with wool from the Great Karoo, passed through the poort for the first time, on their way to Mossel Bay.[5]

One appreciates just how important this connection to the Great Karoo must have been when it is realised that the earliest marine structure in Mossel Bay, Bland's Jetty, which had been built only in 1854, was in 1858 lengthened to cater for the additional shipping demands resulting from the completion of the road through Meiringspoort.[6] By 1870 well over one million kilograms of wool, one-eighth of the wool clip of the entire Cape Colony, was being transported through Meiringspoort for sale at Mossel Bay.

◆◆◆◆◆◆◆◆◆

Unfortunately, as the engineers had advised would be the case, the road was subject to flooding, and passage was often blocked for weeks on end. In November 1859 the road suffered severe flood damage. In December 1861 another flood caused the pass to be closed for a month. The road was completely washed away in October 1875 by a succession of heavy floods, with the road impassable for upwards of a month. A major flood in May 1885 'washed away every vestige of road in Meiringspoort' as well as doing considerable damage to other roads and passes, so that the local situation was most dire until the road was rebuilt in 1886.[7]

Two views of the road, deep in the poort (P. Wagner, courtesy of Photo Access)

These damaging floods are mentioned as illustrations of the precarious nature of this major artery. (It is this vulnerability which justified the construction of the awe-inspiring, but flood-free, Swartberg Pass between 1881 and 1888.) The road through Meiringspoort has been damaged by floods periodically since then, and much ingenuity, hard work and money have been expended in improving the situation. For example, in the 1930s £10,000 of Depression money was spent on the road,[8] and between 1948 and 1953 the Oudtshoorn and Prince Albert Divisional Councils replaced the gravel-and-stone drifts with concrete causeways at a cost of £14,928.[9]

Then, in the 1960s a Cape Provincial Roads construction unit built the National Road from Beaufort West to Klaarstroom and, between 1967 and 1971, reconstructed and surfaced the continuation through Meiringspoort as a trunk road. John de Kock has recorded his recollections of this project for us.[10]

In his monograph John recalls how a decision had to be made between a route tunnelling through the Swartberg and one following the existing surface route via Meiringspoort. The tunnel route would have accommodated a high-standard road but required a series of tunnels and heavy earthworks. Extensive planning and ultimately specialised tunnelling would have been essential, and even the preliminary work would have tied up the Provincial unit for some time. This route had obvious advantages, especially as regards removing the risk of flood damage to the road, but it would have been extremely expensive.

The poort route had obvious disadvantages: because of the sinuosity of the poort (to name but one factor) the road could not be built to National Road design standards, and there remained the ever-present threat of damage by floods. It would, however, be much cheaper, enabling more of the road funds to be allocated to other needy road sections, and of course the views and other environmental considerations would be superior to those found inside a tunnel.

After considerable discussion it was decided to build the road through the poort as a trunk road. Construction started during 1967.

Environmental factors played a major role in planning the work, as it was rightly considered that the poort had to be preserved and its character maintained. Among other things, this meant that blasting, and the subsequent exposure of raw rock faces, would be limited as far as possible. This resulted in an increase in the use of retaining walls, which either were built of natural stone or, if of concrete, were faced with stone to blend into the surroundings.

The stream through the poort had to be crossed 25 times. Obviously bridges of a size to accommodate the flood run-off from the catchment area in the Great Karoo would be too intrusive, so it was decided to provide the crossings with a nominal opening to take normal flow, with protective walling and paving to accommodate flow over the slabs. These were not designed as low-water bridges but as road slabs so that the flow would not be concentrated. It was accepted that the road could be over-topped during some floods, but the river is subject to rapid run-off, as is the nature of Karoo weather, so any resulting closures would be of short duration. Extensive protection works were incorporated into the design to mitigate damage caused by such floods.

Very detailed surveying and checking were necessary, but John de Kock wrote that 'even so, the final alignment had to be decided on site, considering rock faces, stone walls, plant growth and flood protection requirements. 'It is necessary to emphasise the importance of "fitting" the road to the poort with constant checks on alignment ... The poort width varies greatly, and in one section road and river are between vertical rock faces where the only solution was extensive concrete works since the road would be part of the streambed during floods.'

This was a mammoth task requiring delicate, almost day-by-day treatment by designers and builders. It was handled in a most environmentally sensitive way by the caring Cape *padmakers*, at a cost of only R1.6 million. The official opening took place on 3 August 1971.[11]

♦♦♦♦♦♦♦♦♦

Although it was an immensely successful project, the road still had to run through the poort, and mostly on ground level, so that it remained vulnerable to major floods. Thus, in early 1974, when general heavy rains caused damage to roads in many parts of the country, a local cloudburst resulted in a flood which caused extensive damage in Meiringspoort.[12]

And so things went on, with flood damage being repaired as it occurred, to keep this important ground-level connection between the Little and Great Karoos open, most of the time. Then a big flood came, in November 1996, and a major redesign and reconstruction had to be carried out.

This 1996 flood severely damaged the road (some reports say the road was 'destroyed', which was probably not an exaggeration) and the poort was effectively closed to traffic for many months.

Reconstruction was carried out during the next three years, with the poort open to traffic – although one-way in sections – whenever possible. It was extremely difficult to work in such a confining corridor, under traffic, and to be delayed several times by repeat flooding. However, by December 1999 it was possible to fully re-open the road to traffic while the finishing touches were being done. Most unfortunately, another flood in March 2000 added R7 million to the cost, which brought the total up to R70 million and delayed the official opening for six months.[13]

Some detail of this reconstruction was given in an article by Tony Murray.[14] 'Following devastating floods ... emergency repairs were carried out by the Provincial Administration of the Western Cape. The permanent repair work was implemented through a two-and-a-half-year contract completed in October 2000. Contracting was a joint venture between Group 5 and Grinaker, and the engineering consultants were Africon Engineering International, who also managed the contract administration, and Liebenberg & Stander.

'The work included repairs to and reconstruction of the 25 low-water bridges in Meiringspoort, construction of retaining walls to protect the roadway, and the reconstruction of approximately 8 km of the 20 km road section. Finally, the section of road was resealed. During construction traffic was accommodated in half-widths and maintained throughout the construction period by 24-hour traffic light control.

'To maintain the pre-flood atmosphere of beauty and masonry stone walls, most flood-protection retaining structures were built by teams of local people using stone obtained from approved sources. Recruitment of labour was also from local communities ...

The gravel road and its 1971 replacement (John A. de Kock)

'The road section falls within the Swartberg Nature Reserve, which is under the control of the Department of Nature Conservation. Various unique flora and fauna species are found in this sensitive poort and strict control measures were imposed [with] an environmental control officer ... to ensure that the environment and beauty of the area were not disturbed by construction activities.

'The Department's Roads Branch also developed the tourism attraction by upgrading a number of rest areas with neat braais, paved parking areas and ablution facilities ...'

It was a demanding job well done. On 20 October 2000 Meiringspoort was officially reopened after being under repair since November 1996. The ceremony was jointly performed by the Premier of the Western Cape, Gerald Morkel, and his Minister of Transport, Piet Meyer. The final result of this refurbishment is most attractive indeed, and I can recommend a leisurely visit to the poort when you are next in the vicinity.

◆◆◆◆◆◆◆◆◆

A spokesman for the Western Province is reported as saying that the road 'could now withstand further flooding'.[15] I sincerely hope so, but looking at the history of Meiringspoort I would have been chary of going on record with such a statement.

CHAPTER FIFTEEN

Seweweekspoort & Bosluiskloof Pass

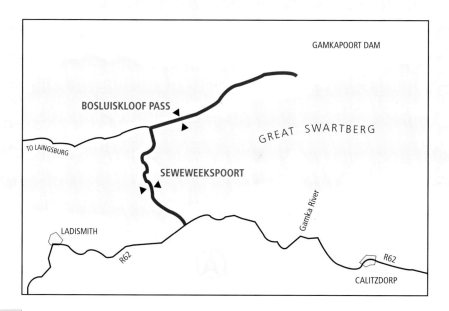

The local inhabitants on both sides of the Swartberg had been agitating for the construction of access roads through the mountain range since about 1845. John Montagu, the Colonial Secretary, visited the area in 1849, and after various investigations it was decided to build first the road through Meiringspoort, and then that through Seweweekspoort, 20 kilometres east of Ladismith.

SEWEWEEKSPOORT

Meiringspoort was completed in 1858, and attention then turned to Seweweekspoort, which was already being used as an access to and from the north, even though it took up to six days' hard travel to struggle through the short poort.

The following year Mr Woodifield completed a survey, and construction started, also in 1859, with 108 convicts under their head overseer, Mr Aspey. Mr Aspey fortunately had sufficient road-making experience to carry on with the preliminary works until Adam de Smidt, who was repairing damage to Meiringspoort caused by the severe flood of November 1859, took over from him in 1860.

The pass is 17 kilometres long, following the course of the Huis River through the mountains. One theory about the origin of the name of Seweweekspoort is that it is named after a

Seweweekspoort sketched by Sophia Gray (NLSA 14314)

Berlin Mission Society preacher Louis Zerwick, who did his good works in the vicinity.¹ Had it not been so named we might have ended up with a Huis River Poort, as well as the Huis River Pass further downstream on the road to Calitzdorp.

Here, as in Meiringspoort, the *padmakers* were faced with a refusal by the Road Board to allocate sufficient funds for a reasonable standard of road. The 'boer road', which was all that could be built along the valley bottom, came under water at the frequent river crossings, often blocking traffic, when the water level in the infant river rose. Severe flooding in 1875 and on other occasions resulted in Seweweekspoort, as well as Meiringspoort, being closed to traffic for some time.²

Construction was sufficiently advanced by June 1862 for the pass to be opened to traffic and it was finally completed five months later. Although susceptible to flood damage, it was a tremendous benefit to the districts which it served, opening up the lines of communication on which economic development is always dependent. A four to six days' journey had been reduced to one of three hours.³

Many have described the rugged beauty of Seweweekspoort, but I shall quote Dr William Atherstone, a much-respected geologist of that period (he it was who identified the first diamond found near Hopetown in 1867). He drove through Seweweekspoort in 1871 with Thomas Bain, also respected as a geologist, when on their way to investigate reports of gold having been found near Prince Albert. Dr Atherstone wrote about

> the most wonderful gorge or mountain pass I have ever beheld. For twelve miles you travel bare walls of vertical rock, in parts 3,000 feet high, twisting and twining as the mountain stream winds through the flexures and curves of the mountain chasm, crossing and recrossing, I am told, more than thirty times; in parts so narrow there is scarcely any

room for the river and road – yet an excellent wagon road has been made through it with comparatively little expense; and, certainly, nowhere in the Colony have I seen so wonderful a pass – a clean zigzag cut through the whole thickness of the rock formation of the range from top to bottom. When once you enter, no appearance of exit is there for two hours and a half; but you are constantly meeting new scenes, over which quartzose cliffs, curved and fractured in every direction – now red vertical sandstones, with flexures and arches jammed together in inexplicable confusion, as if jammed together laterally by prodigious force – at the next turning, gentle ripple-like rock waves, with blue slate – and high overhead, bright-yellow lichened crags, making the neck ache in an attempt to look up at them, with a small chink of sky over head; shut up in front and behind, with the white river-bed below, or on one side curved with huge quartz boulders, and fringed with green trees – keurboom and wagenboom, aloes, and succulents nestling in the rock-fissures high above you. How few know of this extraordinary mountain gap![4]

All these remarks – even the last sentence – apply today, 130 years later: the road still has a gravel surface.

BOSLUISKLOOF PASS

Dr Atherstone travelled by post cart from Grahamstown to George, where he was joined by Thomas Bain. They then proceeded by cart via Riversdale, Plattekop (Platteklip) Pass, Ladismith, Seweweekspoort and Bosluis Pass to Prince Albert. Look at a map and you will probably be as puzzled as I am as to why they did not travel via Montagu Pass and

Bosluiskloof, sketched by Sophia Gray in 1869 (NLSA 14315)

Seweweekspoort (David Steele, courtesy of Photo Access)

Meiringspoort, both of which had been constructed by that time. Be that as it may, we have not only been left with the description of Seweweekspoort but he also tells of their finds in Bosluiskloof.

Bosluiskloof is an attractive little pass, to the east of the northern end of Seweweekspoort, which was left stranded, as it were, when the construction of the Gamkapoort Dam in 1968 cut off the continuation of the road to Prince Albert. The pass has been termed 'the gateway to the Gouph', an area on the Gamka River noted for its fertility. Adam de Smidt built Bosluiskloof Pass while and after working on Seweweekspoort, to complete the natural link from Prince Albert to the west.[5] It was not particularly heavy construction, but the pass has a wild beauty of its own. Atherstone again:

> ... a scene burst upon us I shall not forget in a hurry. Breathless I gazed down the valley on the boundless sea of blue mountains, cones and peaks, table tops and jagged lines of hillocks, tingled with the faint blush of early morn – the huddled groups of hills in the mid-distance, still in deep shadow; with the aloes and crassulas, and the fantastic rocks of the Zwartberg on our right, and the road winding down the steep sides of the Little

Zwartberg, whose topmost crags were just painted by the glowing rays of the unseen sun. What a wild charm thrown over the distant labyrinth of hills in the soft glow of early morn!

The rocks of the pass contain fossils of 'creatures before the flood', the most prolific being fossils of (bush) ticks – hence the name of the river and the pass. In Atherstone's day there was an abundance of various fossils, but I fear that they are not as obvious nowadays.

From a point about five kilometres west of the dam the narrow footpath to Gamkaskloof (previously sometimes known as 'Die Hel') takes off. One section of this footpath is known as 'Die Leer': here it zigzags down an almost vertical drop of about 450 metres. Only for the strong of heart and nimble of foot – and a clear conscience would also be an advantage!

◆◆◆◆◆◆◆◆◆◆

With the N1 charging up through Laingsburg, and the direct Cape Town–Oudtshoorn route via Ladismith being provided with an excellent permanent surface, the gravel road through Seweweekspoort is little used by the hurrying, scurrying working travellers of today. The locals use it (although the Gamkapoort Dam has cut this number), and of course it is one of a variety of options for tourists and for those interested in our mountain passes.

So may I suggest that the next time you travel between Cape Town and Oudtshoorn you think about going via Laingsburg and Seweweekspoort? Allow a little extra time, because you won't want to zip through the poort – you will want to stop now and then and get out so that you can gaze upwards at the marvellous colours and shapes of the rocks. The poort, cutting through the mountain as it does, provides one with a quite unique visual physical cross-section of the geology, which is usually hidden deep beneath the surface of a mountain.

And the time after that (because you will want to do it again), allow a couple of extra hours so that you can also pop down Bosluiskloof and back again.

So much to see, and so little time to see it in.

Mountain scenery in Seweweekspoort (Kelvin Saunders)

CHAPTER SIXTEEN

THE MESSELPAD PASS

The people of Namaqualand led a pleasant, relaxed pastoral and largely nomadic existence until copper deposits in the Springbok–Okiep–Concordia area began to be worked in the 1850s. Pretty soon copper was a dominating consideration in the lives of the Namaqualanders, both as regards winning the ore from the ground, and also as regards transporting the ore to the coast for export to Britain for refining. By 1860 copper – all from Namaqualand – had become the Cape Colony's second most important export.[1]

The development of transportation facilities to meet the growing demands of the mines is a story in itself,[2] but all links to the ports had one factor in common: they had to traverse the mountains of the Hardeveld (in which the mineral deposits were located) and descend their western escarpment before struggling across the soft 'beach' of the Sandveld for the final third of their journey.

Initially the ore was exported through Hondeklip Bay, lying 125 kilometres (or six days by ox wagon) southwest of Springbok. To reach this anchorage the wagons of the copper riders had to cross the mountains between the Buffels River and Wildepaardehoek on the fringe of the Sandveld. This chapter is the story of the development of what came to be known as the Messelpad Pass through Tiger Kloof.

The mining of copper promised Namaqualand an economic boom of proportions never before known in the region. It soon became apparent, however, that unless the problems associated with getting the ore to a port, and in transporting coke and other supplies from the port to the mines, could be overcome, this economic balloon would expire without trace. Of the total cost of £16 to transport a ton of ore from the mines via Hondeklip Bay and London to Swansea for refining, £10 10s was the cost of land transport to Hondeklip Bay.[3] Once again, the truth that all economic development is dependent on transportation was being forced to the notice of those concerned.

Transport was at first, logically enough, by ox wagon. Initially these copper riders were the local farmers, of various ethnic groups, and they used their trek wagons and trek oxen for this purpose. It was a great opportunity to get some cash money, but the trek wagons were really rather light for the job, the terrain they traversed was, to say the least, unfriendly, and many who set out with 1,400 kilograms of bagged ore arrived with only 900 kilograms of freight.[4]

This copper riding was a hard and rugged exercise. An ox wagon would take six days from Okiep to Hondeklip Bay and four days on the return journey: a mule wagon would take eight days for the round trip. By 1866 there were over 300 wagons and 6,000 draught animals on the Hondeklip Bay road at the same time.

The original 'copper road' scrambled for 30 kilometres over precipitous brows and mountain ridges, with little or no constructed formation. The most demanding section was that west of the Buffels River crossing, where they had to climb over the final barrier and descend from the escarpment to the flattish Sandveld section at Wildepaardehoek. This section was particularly wicked on man and beast, and added greatly to the cost of transporting the copper ore to the coast.

◆◆◆◆◆◆◆◆◆

The people of Namaqualand, and especially the Magistrate and Civil Commissioner, Joshua Rivers, peppered the colonial government with memorials and deputations requesting that something be done to improve the access from the mines to Hondeklip Bay.[5] The Divisional Council reported that their funds were inadequate even to keep the existing copper route in repair, let alone to finance any construction works. And all this time the copper companies were petitioning the government to build a road or a railway, or both, to enable them to get their ore to a port.[6]

It therefore comes as no great surprise to find that the government, after the normal lag period, called for reports on the overland transportation problems, and for recommendations as to how these might be overcome. The first to put in his report was Andrew Geddes Bain, who presented to both Houses of Parliament the first geological report dealing with the Namaqualand copper mines. Regarding copper riding, his words have often been quoted: 'Without some grand improvements in the roads, the Mining Companies never can advance. Hundreds of tons of ore are now lying at the different mines, which the proprietors cannot get conveyed to the coast, at any price.'⁷ Reports were also submitted by Commander M.S. Nolloth of the survey vessel HMS *Frolic*, by Charles Bell, the Surveyor-General, and by Dr Andrew Wyley, the government geologist,⁸ and motivated requests for assistance continued to arrive from the locals.

In 1865 the Cape Copper Company appointed the civil engineer Thomas Hall to investigate independently and report on the transport of ore from the mines to the coast. He teamed up with Patrick Fletcher, the government surveyor for the district, who knew the country well. After six months' concentrated work, covering over 2,000 kilometres of country and levelling and measuring 120 kilometres of routes, he and Fletcher submitted their findings to the Colonial Secretary and the Chief Inspector of Roads (who accepted the recommendations) and to the Governor (who said he had no funds).

Hall then prepared and submitted his report to the Copper Company. Although his favoured proposal was to construct a narrow-gauge railway from Port Nolloth (150 kilometres north of Hondeklip Bay) to the mines, he also of course investigated the route over the escarpment to Hondeklip Bay. Here his recommendation was to abandon the route in use,

The Messelpad Pass (Graham Ross)

which crossed the mountain summits to avoid the wagons capsizing on a cross-slope, and to construct a new route along the side of Tiger Kloof. Such a demanding construction was of course beyond the capabilities of the copper riders or the copper mines, and required government funding.[9]

Finally, in January 1867, we see some action on the part of the colonial government, but only after the promise of local assistance. The Cape Copper Company undertook to provide transport and accommodation for the convicts to be employed on the construction of the road to Hondeklip Bay. In addition they promised to pay two annual instalments of £800 each to cover further establishment costs. (In actual fact, following a petition in 1868 the Company was released from payment of the £1,600, in appreciation of the fact that they had already at that time spent £1,835 on transport and accommodation.)

◆◆◆◆◆◆◆◆◆◆

The surveyor Patrick Fletcher, who was regraded as Road Inspector, falling under the Chief Inspector of Roads, M.R. Robinson, was placed in charge of the construction of the new

A fine example of dry-stone retaining wall (Graham Ross)

route over the mountain. This is today known as the Messelpad because of the dramatic dry-stone retaining walls through Tiger Kloof.

It is interesting to learn that when Patrick Fletcher landed in Cape Town circa 1850, he found that there was practically no opening in the young Colony for civil engineers or geologists. So he studied, and passed the necessary examinations, and was admitted to practise as a land surveyor. He had now returned to his 'first love', civil engineering.

As this was without doubt the greatest pass project undertaken in Namaqualand in the nineteenth century, I shall give more detail than for most of the other passes covered in this book, to convey a clearer picture of the trials and tribulations involved.

Fletcher was appointed to the road project from 1 January 1867. His first batch of convicts landed at Hondeklip Bay on the 8th, and during the year 284 convicts, under Superintendent Pitt, arrived on the job. The first six months were occupied in erecting barracks, quarters, a smithy, cook house and storeroom, while Fletcher set out the road. With 284 convicts plus other staff on site, the supply of water was critical. Various pits were dug and dams constructed to this end and also to provide water for the copper riders, for, as Fletcher truly said, at that time 'a good road without drink-water for cattle is comparatively useless'. The plan he adopted was to do away with culverts as much as possible, providing a spillway for any water the roadworks dam could not contain.[10] (It is interesting to note that the Resident Engineer in charge of the National Road construction near Garies used the same technique a century later, in 1966.)[11]

The terrain through which Fletcher had to drive his road to avoid the steep and hazardous route across the mountain tops was hard and unfriendly, necessitating much blasting and manhandling of rock. The work tackled in the first year (1867) opened up a route for vehicles from near the main convict station (which was where the route left the Buffels River) to the top of Tiger Kloof. Combined with a natural track in the bed of the Buffels River eastwards to the toll house, this provided a 12-kilometre length of roadway with flat gradients, which enabled traffic to avoid the old route over the mountain tops between these two points. Fletcher's report includes a contour plan, which shows what a welcome improvement the new route must have been. The first cart went over the new road on 28 October 1867. There is a rather delightful conclusion to his report where he politely asks the Chief Inspector to avoid sending him 'totally inexperienced road overseers' in future.

During the second year (1868) the work was confined to the section between Wildepaardehoek and the ascent on the northern side of the Buffels River. A minimum width of ten feet (or three metres) was blasted out for a further eight kilometres from the end of previous construction at the top of Tiger Kloof. Fletcher also surveyed and reported on the materials along the five sections of road between the Bay and Springbok.[12]

Fletcher still worked on the tricky eight-kilometre section from the top of Tiger Kloof to the main convict station during the third year (1869), widening the roadway now to 18 feet so that the section between Tiger Kloof and the Buffels River drift, a distance of 14 kilometres, was all but completed.[13]

The completion of this construction as originally planned was now overtaken by other events. As the output of the mines had increased, so had the demand for transport to the coast. With this increase in activity the copper riders' stock was eating all the agricultural produce of the region, as well as large amounts of additional supplies imported from Malmesbury (500 kilometres to the south), and large quantities of grass imported from the plains of Bushmanland to the east.[14]

Then came drought years, and this setback was aggravated by lung-sickness, from which hundreds of draught oxen died. There had been a serious suggestion in 1857 from the government geologist, Andrew Wyley, that camels should be used, as 'three camels would carry a ton weight from the mines to Hondeklip Bay or Robbe Bay [Port Nolloth] in three days – a task now requiring ten mules'. Although proven by use in other places, this suggestion does not appear to have been followed up in Namaqualand. In time mules proved more suitable than oxen, particularly over the mountain sections, and mules began to replace oxen in the 1860s; by 1871 most of the draught animals were mules.[15]

There remained the fact that transport by wagon was restricted to a short riding season. The rain, what there is of it, falls mainly in winter and grass follows on but withers and dies in the summer heat. Water along the riding routes is scarce and brack. So climatic conditions dictated that there was only a short riding season while fodder was available, and at other times there was no way for the miners to get their product to the port.[16]

It was obviously uneconomic for the companies to have stockpiles of ore lying at the mines, awaiting transport to the coast, and it had become very apparent that transport by wagon was unable to keep up with the increasing mining production. Towards the end of 1868 the Cape Copper Company decided that the situation was becoming critical, and they called Thomas Hall back to Namaqualand to construct the Port Nolloth–Okiep railroad he had recommended in 1866.

This decision of the copper companies naturally reduced the pressure to complete the Messelpad deviation, not to mention the need to allocate finance to construct the remainder of the Hondeklip Bay road. From then on, Fletcher battled against a shortage of funds. In January 1870 he proposed reducing standards over various sections of his project, which, while enabling all the hills on the south side of the Buffels River to be avoided (thus turning the portion already constructed to best advantage), would also remove the principal barriers on the north side of the river.[17]

During 1870 and the first three months of 1871 Fletcher worked on the abbreviated project to the reduced standards specified. For a man who obviously had pride in his work, this was depressing. He managed to make Tiger Kloof passable but had to use inferior material, which the wagons ploughed up: his completion report pleaded for an additional allocation of funds to enable the surface to be improved. He had an average of only 77 convicts during 1870 and these were all removed at the end of the year. A gang of 20 'free labourers, natives from Kamaggus', who 'upon the whole gave satisfaction', was recruited in June 1870 and worked through to the end of the job in March 1871. In the same month the Divisional Council of Namaqualand took over responsibility for the maintenance of the road, and also took over his tools for a nominal £50.

The worst mountain section of the Hondeklip Bay road had finally been improved. Patrick Fletcher had constructed 14 kilometres of road five and a half metres wide involving very heavy work through Tiger Kloof, besides improving a considerable length of approach roads, through some of the toughest country in the region. The original estimate for the whole 28-kilometre section between Wildepaardehoek and Jakhals Water was £5,000: his total direct cost was £4,025, but this did not make any allowance for the average of 150 convicts engaged on the work. It has been suggested that these might be costed at 2s 6d per day – which of course considerably increases the cost of the road.

After June 1870, when Thomas Hall had brought his tramway from Port Nolloth across the heavy sand to Abbevlaack, the emphasis turned from riding copper ore to Hondeklip Bay to riding it to the railhead. This switch increased as the railhead advanced further and further towards Okiep, and when the steel reached the mines in January 1876 the era of the copper riders transporting ore by wagon to Hondeklip Bay was over.

◆◆◆◆◆◆◆◆◆

The construction of Messelpad Pass, making a good wagon road along the steep sides of Tiger Kloof without any of the modern aids which present-day road-builders take for granted, is truly worthy of the National Monument status which has been conferred upon it. The dry-stone walls which Fletcher constructed on the project compare very favourably with those by Thomas Bain on some of the better-known mountain passes further south.

Fletcher's pass is worthy of far greater general appreciation than has been its lot. His work stands pretty well in its original form to this day. A study of any road map of Namaqualand will demonstrate the strategic importance of the route: the direct link between Hondeklip Bay and the division's administrative capital, Springbok. The road is in constant use.

Good maintenance by the Divisional Council *padmakers* has of course been a major contribution to this success, but the fact that no major realignment or upgrading of geometrics has been found necessary speaks volumes for the standard of Fletcher's design and construction.

CHAPTER SEVENTEEN

COGMANS KLOOF PASS

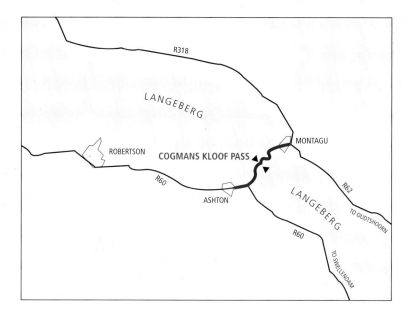

C ogmans Kloof Pass traverses a poort through the Langeberg between Ashton and Montagu, on the road to Oudtshoorn. At the time it was completed the nearest alternative passes through these mountain ranges were Hex River Pass, 50 kilometres to the west, and Tradouw Pass, the same distance to the east.

The pass follows the course of the Kingna River (sometimes known as the Cogmans River) and, as with Meiringspoort and Seweweekspoort described earlier, is thus liable to damage by floods. Luckily, the valley bottom of the Kingna is reasonably wide, so that it has been possible for most of its length to construct the road clear of the riverbed and thus above the level of 'normal' flooding.

The Cogmans, Kogmans or Kockemans (the spelling varies) were a Khoikhoi chiefdom that attacked the Land of Waveren military outpost in 1701. I was unable to find out who won the skirmish, but to this day the kloof is named after the clan – not after a Mr Cogman, as some people may imagine.[1]

The early history of white settlement in this region is not well known though farms were granted beyond the Langeberg from 1725 onwards. But their most direct connection to the mother settlement was rather precarious. The original track through the kloof more or less followed the riverbed, with eight hazardous drifts, some of heavy sand and others over rough boulder beds, and to round the Kalkoenkrantz wagons had to travel in the stream bed itself as there was no other way through. In the early 1850s this first track was somewhat improved by the Swellendam Divisional Council, but it was easily washed away by floods, often leaving it impassable for a month or more on end. It really defied all efforts at keeping it in repair.²

Then, as now, the almost automatic reaction to a cry by citizens for assistance was to appoint a committee. Thus we find that a parliamentary Select Committee decided in 1861 that a road should be built, using convict labour. But as no convicts were available at that time, it was of course not possible to actually do anything.

It took the loss of twelve lives by drowning in a flood in 1867 to get some action. A flying survey of the pass was undertaken in 1869 by Mr M.R. Robinson, then Chief Inspector of Roads. An estimate of £10,000 (reduced to £2,500 if convict labour was used) for five and a half kilometres of roadwork was approved, and construction started using 'distressed labourers'. But unfortunately this effort ran out of steam – these previously unemployed labourers were not physically capable of the back-breaking work required – and work stopped in 1870.

However, they must have accomplished something, because we find in the *Standard and Mail* of 20 January 1872 a report of the opening ten days earlier of the first section of the pass by Civil Commissioner Hodges of Robertson. His daughter was on the same occasion called upon to christen and name the 'Hodges Bridge'. The newspaper report mentions that 'the only accident that happened was the bursting of one of the cannons when the salute was fired on the return from the opening of the Kloof, but that fortunately did no damage'. This jolly occasion, prob-

Bain's 1877 road along the Kingna River (NLSA 19418)

ably organised to encourage the authorities to recommence construction, included 'tiffin' in tents on the river bank where a considerable number of toasts were drunk to all and sundry, and a croquet match between teams from Robertson and Montagu.

In due course Thomas Bain was called to the rescue, as we see happening also in other passes where progress had ground to a halt. Bain surveyed the kloof in 1872 and set out the

Mountains upended in some titanic struggle (David Steele, courtesy of Photo Access)

work, and C. Hendy (sometimes called Bill Handey or Henley) started construction with 32 labourers – all they could enrol, as the hard work was not overly attractive – in 1873. Hendy had been with Thomas Bain on his (Bain's) first pass-building job, Grey's Pass near Citrusdal, and has been referred to as Thomas's 'staunch and efficient Mr C. Hendy ... his excellent chief overseer'.[3] He had also been in charge of the previous 'distressed labourer' effort. With such a stalwart on the job and with Thomas Bain to keep an eye on things, it is no wonder that this construction was carried through to a successful conclusion.

The job was five and a half kilometres long, and included the well-known unlined tunnel through hard rock under the Kalkoenkrantz, initially (it was enlarged later) 16 metres long, with a five-metre-high arched roof, which avoided two dangerous drifts. This tunnel was a brave work to be tackled at that time, especially when it is recalled that Thomas's father, Andrew Geddes Bain, had been unsuccessful in his attempt to construct the first road tunnel in the country in Bain's Kloof 25 years previously.

Remember also that dynamite, with its disruptive force about eight times that of gunpowder, had only been discovered in 1867, and although Bain did manage to obtain some of this new explosive, supplies ran out and most of the blasting had to be done with the inferior, and much more dangerous, gunpowder.[4]

Immediately at the downstream or southern end of the tunnel the road twisted sharply to the left. In fact the tunnel itself was curved at this end. The road was then supported against the slope of the mountain by a dry-stone retaining wall along the left river bank – which can still be seen today.

Work was carried out from 1873 to 1877 at a cost of £12,282 3s 10d (I love these precise figures). Magistrate Hodges was once again called upon to do his thing, and performed the opening ceremony on 28 February 1877, with his daughter on this occasion christening 'Bain's Tunnel'.[5]

The farmers to the east had at last, 150 years after the first farms were given out, obtained their all-weather connection with the markets to the west, or nearly so! True, the really difficult section through the kloof and the Kalkoenkrantz had been improved, but parliamentary approval had been limited to the construction of this five-and-a-half-kilometre section. Access was still liable to be cut off after heavy rains where the road crossed the Kingna River at the western entrance to Montagu village. This situation was in fact only relieved when the De Waal Bridge was constructed here in 1915.[6]

The tunnel through the Kalkoenkrantz (Graham Ross)

◆◆◆◆◆◆◆◆◆

The next stage in this saga was the 'tarring' of Bain's road.[7] As anyone who has lived at the end of a gravel road can vouch, this was a major event for the locals. The Montagu Divisional Council organised a formal luncheon on the occasion of the official opening on Saturday 21 November 1931.

Traffic volume built up over the years, and the size of vehicles in common use also increased. Bain's tunnel, with its sharp turn at the southern exit, needed improvement to cater for traffic demand, and the road generally required upgrading. The alignment was improved by straightening (and widening) the tunnel and building the Loftus Bridge (with three 40-foot spans) to carry the roadway across to the right bank of the river. West-

The retaining wall which protected Bain's road from the 1981 flood (Graham Ross)

bound traffic was brought back to the left bank by the Boy Retief Bridge (four 40-foot spans) further downstream. The road itself through the new pass was constructed to standards 'suitable to meet the demands of modern traffic', as the Provincial Roads Engineer put it in his 1953 Annual Report.[8]

The standards certainly were an improvement on those provided by Bain 75 years earlier. But to attain them it was necessary to cross the river twice, which Bain had avoided. True, all care was taken to ensure that the bridges were designed to cater for normal floods, but on 25 January 1981, the same day that Laingsburg was so tragically flooded, the Montagu Hot Springs complex was destroyed, much damage was done to other property along the river including the road, and the approaches to the Boy Retief Bridge were washed away by a most abnormal flood.

Then were Thomas Bain's words proven to be true. He had reported in 1877: 'The Cogmans Kloof pass has been very substantially made. The walls are massive and well built and the road is protected by good drains and culverts ... the pass will not be subject to much damage by heavy rains and will consequently cost little in repairs.' Until the 1953 road could be repaired Bain's old, abandoned pass was used to provide access to Montagu.[9]

✦✦✦✦✦✦✦✦✦✦

On 6 September 1978, a century and a bit after the construction of the pass by Hendy and Bain, a plaque honouring Thomas Bain was erected by the National Monuments Council in the picnic area on the Montagu side of the tunnel. It reads:

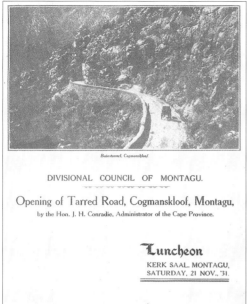

Menu for lunch on the occasion of the opening of the tarred road (Montagu Museum)

This pass was built 1873–1877 by Thomas Charles John Bain (1830–1893), son of the pioneer road-builder Andrew Geddes Bain. Thomas Bain built many roads and twenty-two passes in the Cape Colony.

The plaque was unveiled by Professor J.J. Oberholtzer, Director of the National Monuments Council, who delivered a suitable address, after which all those who had been invited adjourned to the town for a luncheon party.[10]

When next you drive through Cogmans Kloof, stop and climb out of your comfortable motor car, walk to the edge of the road and really study the riverbed. Imagine crossing eight stony or soft sandy drifts. Imagine being left with no option but to take your wagon around the Kalkoenkrantz in the stream bed. And you will agree that it is no wonder that Thomas Bain and his fellow *padmakers* were appreciated in their lifetimes, and are still venerated today.

CHAPTER EIGHTEEN

KOO OR BURGER'S PASS

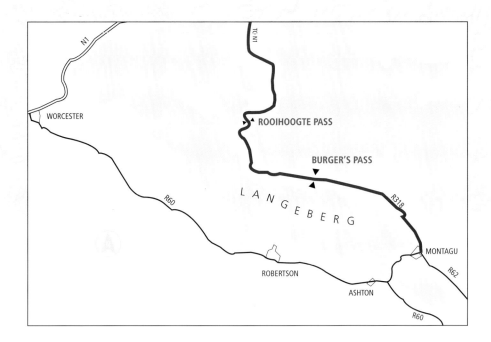

Most of the mountain passes mentioned in this work cross lengthy mountain ranges which lie in the way of the traveller. But these primary routes had themselves to be joined the one to the other by secondary routes parallel to the mountains, to complete the road network and also to give local access to the areas between the primary routes.

This chapter considers the secondary route leaving National Route 1 at Matroosberg, just east of the Hex River Pass summit, to connect up with the R62 in Montagu, at the eastern end of Cogmans Kloof, about 50 kilometres away as the crow flies. There are two passes on this road: Koo Pass, later renamed Burger's Pass, and Rooihoogte.

KOO MOUNTAIN PASS

Thomas Bain was building Cogmans Kloof between 1873 and 1877. As frequently happens, the presence of an engineering work initiated thoughts of other works in the minds of the local inhabitants, and Thomas Bain was called upon to look at providing access to the virtually land-locked but extremely fertile fruit-growing valley known as the Koo. The farmers there were having difficulty in getting their tender produce to the railway for transport to market, and if you look at a topographic map you will readily see why that was so.

As is common in such investigations, local farmers, especially if they were also Divisional Councillors, came along to give the engineer the benefit of their advice and knowledge of the country. Various routes had their supporters and opponents. As Patricia Storrar puts it so ably: 'the route the proposed road should take through the Lower Wittebergen was a matter for heated and prolonged argument. The Footpadsberg, 'a sort of gap to the east of Patatesfontein', would be too steep and costly; the gap to the west of Patatesfontein was objected to for the same reasons – and so on.'[1]

But Bain finally decided on the route for what he decided to call the Koo Mountain Pass, which of necessity also included the adjacent Rooihoogte Pass. He determined the location of the road, and the construction was carried out by the Divisional Council. The project ran out of funds in 1876, but this was soon remedied and the work was completed in the following year, at a total cost of just £1,000. Thomas Bain commented that this was 'very reasonable', as it involved the construction of two passes.[2] Certainly the cost compares favourably with those of other works carried out at that time.

The construction of this road, with all the benefits which it brought to the farmers in the Koo, was undoubtedly a feather in the cap of the Divisional Council. Some of the dry-stone walling can be seen to this day, and reflects the high standard of work which was attained.

As methods of transport improve, the dimensions of road vehicles increase, and the volume of traffic also increases, it is necessary to improve road standards to accommodate these new vehicles. Thus we find that in 1924 a 400-yard deviation was constructed as a relief work (this was the beginning of the Great Depression). The deviation improved one of the most troublesome sections on the original route where gradients of up to 1 in 5 had often rendered the road impassable during the rainy season.[3] Patrick Coyne records that the early Overland cars had to reverse up the

Burger's Pass (Gunther Komnick)

pass as the steepness of the road, combined with the location of the petrol tanks, stopped the (gravity) flow of petrol to the engines.[4]

We now come to pass names – again. Bain referred to the Koo Mountain Pass, but it was also known locally at that time as the Koodoosberg Pass. And, as we shall see shortly, it was re-christened Burger's Pass in 1951. Then, Bain initially refers to the complementary Rooihoogte Pass as Thompson's Pass – although this name does not subsequently feature in the records of the time, as far as I have been able to find out.

BURGER'S PASS

Montagu division was fortunate to have as one of its Divisional Councillors Piet Willem van Hesland Burger. He apparently campaigned strenuously throughout his service (1930–1954) for the improvement of Koo Pass. The Divisional Council was busy from 1943 to 1951 on a five-kilometre section of pure rock work on the pass, completing the construction with a gravel surface. The gradient was improved to 1 in 15 except for one short section of 1 in 13, and curve radii were kept above 300 feet except for one 'unavoidable' curve of 150 feet radius. The cost of this improvement work was £30,000.[5]

Councillor Burger planted blue gums along the road and at picnic spots, and was honoured for his continuing efforts in connection with the road when it was renamed Burger's Pass at the official re-opening on 18 May 1951.

One further item finishes off this short chapter. In 1960 the pass was reconstructed to permanent surface standards, and you now ride over it on bitumen.

Rose Willis has this to say about the R318 link:

> The [Rooihoogte] pass has some very steep curves and a tight hairpin bend, but it offers excellent views of the fruit-growing valley, known as the Koo, below. In spring this is a froth of pink and white fruit blossoms.
>
> The road exits from this short, sharp pass and after a short twist crosses a small plateau before reaching the beautiful views across the valley which are gained from Burger's Pass.
>
> The pass has a large picnic spot at which motorists often enjoy a refreshment as they drink in the scenery of the Koo ... a rugged area north of Montagu which is famous for its apples, pears, peaches and apricots.[6]

This route, little used by tourists, indeed offers a pleasant unhurried drive, whether towing a caravan or running solo, over two passes through very interesting country.

CHAPTER NINETEEN

GROOT RIVER & BLOUKRANS PASSES

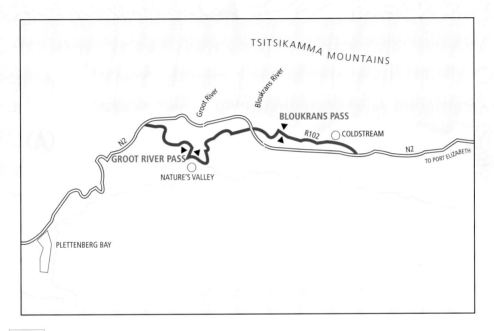

The Great Fire of 1869, which swept through the coastal strip from Riversdale to Uitenhage, enabled Thomas Bain to thrust a road through the previously 'impenetrable' Tsitsikamma forests between Knysna and Humansdorp, opening up a route to the Eastern Province along the coastal shelf. This removed the necessity for such traffic clambering over the Outeniqua Mountains to reach the road down the Langkloof.

Besides all the other trials and tribulations which beset a *padmaker* building 185 kilometres of road through trackless country, Bain also had to take his road through two formidable river gorges east of Plettenberg Bay. These crossings were the Groot River and the Bloukrans Passes.

Many early travellers either tried unsuccessfully to penetrate beyond Plettenberg Bay, or heeded the words of those who had been there, seen that, and routed themselves over Duiwelskop or Prince Alfred's Pass to the Langkloof. Charles Michell wrote in 1839: 'There is no practical way – not even a footpath, from Plettenberg Bay to the Tzitzikamma country where the range dies away. The whole of that extent being intersected with deep ravines and chasms, besides being covered with impenetrable forests.'

The concept of a 'Tsitsikamma Road' along the coastal shelf had been discussed for years before Thomas Bain got instructions to carry out the project. He had collaborated with the

Conservator of Forests, Captain Harison, in 1867 to prepare a report on the control of the indigenous forests.[1] His major contribution to the report was the logical proposal to open up the area by putting a road through it.

Bain's daughter Georgina Lister says that in 1869 'my father … made a survey and map of the almost unexplored Zitzikama Forest country beyond Plettenberg Bay … [he] thoroughly enjoyed his arduous trek. There being no roads he had to cut tracks for his wagons.'[2]

✦✦✦✦✦✦✦✦✦

By 1879 Bain had completed the preliminary location work, and in November of that year his team started the construction of the necessary bridle path to precede the major construction effort. As was (and is) common, work was commenced at different places to suit the needs of access, but the three major gorges, Groot River, Bloukrans and, further east, Storms River, all called for extended construction periods. The road was opened through the Groot River gorge and as far as Bloukrans by 1881. Storms River gorge was reached by 1884,[3] and by 1885 the head of construction was passing Kareedouw. Thereafter Bain's road had only to extend over the level and more open country to meet the Divisional Council road at Humansdorp.[4]

Groot River bridge, at the foot of the pass (Kelvin Saunders)

Bloukrans River bridge (Graham Ross)

Groot River is the westernmost gorge. Bain caused a camp and convict station to be erected there, and the 220-metre descent and re-ascent was completed in early 1882. The actual pass sections are seven kilometres long, but there is in addition the flat section at the bottom past Nature's Valley.[5] Bloukrans Pass, ten kilometres further on over the Platbos plateau, was acknowledged from early in the planning to be the most difficult section of the whole road, and Bain sited his main station there. The pass itself required some deep scarping, and a number of quite high retaining walls along its seven-kilometre length.[6] Work here was completed in 1883. Compared to these two, the Storms River pass was almost insignificant: it was finished in 1885. After all that toil and trouble it must have galled Thomas to have the Chief Inspector refer to his project as being 'a somewhat shorter and more cheerful route than that through the Long Kloof'.

✦✦✦✦✦✦✦✦✦

This route was part of the first National Road scheme declared in 1936, but nothing much happened until after the hiatus caused by the *padmakers* going away to the 1939–45 war.

John de Kock, Resident Engineer, District Roads Engineer and finally the Provincial Roads Engineer, prepared a monograph on the subsequent history of the Groot River and Bloukrans Passes.[7] He wrote:

'In 1946, when the reconstruction of the National Road network got under way with the formation of Provincial construction units, it was decided to place a unit at Keurbooms

Typical of the passes: curves and trees (Graham Ross)

River for the construction of the Keurbooms River–Storms River section. During 1948 the bridges in the Bloukrans Pass were widened ... Both Groot River and Bloukrans Passes were originally built along the – probably – most acceptable line, avoiding extensive cuts and fills, and generally in the shale band. To obtain a "modern" line would have required heavy earthworks and caused extensive destruction of the forest area, and the decision was of course to stick to the old (gravel) road, widen judiciously and follow a "make safe" policy ...

'The alignment of new roads was receiving much attention at that time – the Storms River bridge line is an example – and at one stage early in 1950 consideration was given by the National Transport Commission to a coastal road from Keurbooms River towards the Tsitsikamma area ... It was thus clear that the two passes, and that line of road, would be a temporary phase in the road development.

'The geology of the area is interesting, with the road in the passes mostly in the shale. The area was unstable and had always been subject to slides from both the old cut face and from fills. The steep natural slopes would also be made unstable by cutting without retaining walls. There were some sections of the old road where the curvature resulted in truck-trailer units having to take up the full width of roadway – the procedure was that the truck would approach the curve slowly, hooting to warn vehicles approaching from the other side, and on occasion the driver's helper would run ahead to check the curve for approaching vehicles.

'The restricted sections were generally widened by trimming the cut slopes, and the still-famous curve in Bloukrans Pass, being steep and about 135 degrees, caused many a hold-up. It must be remembered that in those years big horse and trailer units were the exception and all such units used the route through the Langkloof. Traffic volumes were of course low.

'The construction procedures for work done in the passes were as follows:
- All work had to be done under traffic, with limited closing of the road.
- For the final layers and the bitumen surfacing, the road was closed for several hours at a time. Base-course quarries (sandstone) were located out of the passes.
- Drainage was a major item. Side drains were generally surfaced and shoulders were not provided – roads were two lanes wide. The area has a high rainfall, resulting in major problems in placing sub-base and base-course. Base-course, in both passes, sometimes under conditions of all-day shade, required careful control. Priming of the base under adverse conditions was sometimes necessary, with the prime being sanded and opened to traffic almost immediately. Surfacing was done in short sections – the road was closed and the complete operation done (two-coat surface treatment).
- All spoil material had to be transported out of the passes and machines had to occupy road space. It must also be remembered that many construction plant items were old or ex-army: some tractors, many trucks – who can forget the army 4x4 trucks? – and others. Hydraulics were still in the future and all equipment was cable-controlled.

A rare level stretch of road (Graham Ross)

'With the building of the Groot River bridge, in 1950–1, to replace the then existing concrete causeway (which had itself replaced the very old stone-packed causeway), the decision was taken to build the bridge on top of the concrete causeway – the new bridge was a low-level design, with collapsible side-wires – in order to avoid having to build new approaches through the indigenous forest. A bypass was made immediately upstream and by today's standards would never have been accepted, but drivers were a hardy breed then and did anything for a surfaced road! Also of interest is that a large stinkwood log was found under this bypass, but the Forestry Department – regrettably – claimed it from the Resident Engineer.

'Safety precautions through the passes consisted, generally, of wooden posts with steel cables. Some sections were provided with timber barriers, but these measures were more in the nature of guide rails than guard rails. The low traffic volumes – and slower-moving vehicles – had few problems. Steel guard rails were not readily available in those years.

'These passes have always required full-time patrols for minor maintenance and will no doubt continue to do so. The shales generally weather badly and several serious slides have occurred, requiring walling. Trees growing above cut slopes have tended to become top-heavy and in wet, windy conditions have fallen over, bringing stone and soil down. Disturbances by baboons have caused minor problems.

'The work through the passes was completed in 1951 and the road remained in use, as a major route, until 1984 when the new N2 toll road was opened. Despite the extra time to traverse the passes, it is still a route to be enjoyed at leisure.

'On a more personal note, I well remember an incident in Bloukrans Pass. I had been setting out the new road line from the top of Bloukrans Pass in the direction of Storms River. It was a Saturday morning and at 1 p.m. we packed up to go back to Keurbooms River camp. The vehicle was a Ford, circa 1941, and I had three survey helpers. The approach into the pass from that side is fairly easy, and as I was about halfway along this downgrade I gently applied the brakes – and yes, no brakes! Having spent some time driving army ammunition trucks in the mountains of Italy, I engaged in some rapid gear changing and with the hand brake – which worked 50 per cent – brought the van to a stop at the first short curve. When I looked back I saw my survey team of three way back in the road – they had jumped off. When I asked them why they had jumped so smartly, they said: 'Meneer, die *van* het al van tevore so gemaak!' I had a word or two with our mechanics.'

The National Road now goes along the plateau, hurdling minor and major chasms as it comes to them.

Storms River was bridged in 1956, the bridge's size and unusual design arousing much interest in engineers and laymen alike. The photographs and descriptions displayed in the tearoom overlooking the bridge give an excellent idea of the methods employed to build and position the arch. As a matter of interest, the opening of this bridge completed the all-asphalt N2 link between Cape Town and Port Elizabeth.

Almost a century after Thomas Bain and his band of helpers triumphantly completed their Tsitsikamma Road, Groot River and Bloukrans gorges were also bridged at the plateau level. This took place in 1983 as part of the construction of the high-standard N2 toll road. The graceful Bloukrans Bridge, 216 metres above the bed of the gorge, is the longest single-span bridge in Africa.

✦✦✦✦✦✦✦✦✦✦✦

By all means travel along the toll road if you are in a really tearing hurry. But if you are not pushed for time, you will enjoy travelling the Bain–De Kock passes much more. It will take you twenty minutes' more driving time. It is up to you how much more time you spend out of your car looking at details of the construction and at the natural beauties and indigenous trees and foliage. There I cannot help you.

CHAPTER TWENTY

THE PASSES ROAD

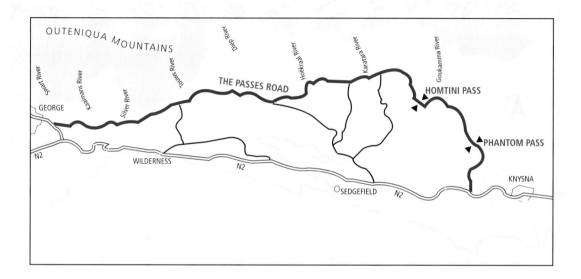

This 75-kilometre road, providing a properly engineered road between George and Knysna, was completed in 1883. Skirting the foothills of the Outeniqua Mountains on the sea side, it has an entirely different character from the National Road which John de Kock built in 1952 between the lakes and the sea. Because the rivers generally flow in deep gorges across the line of road on their way from the mountains to the sea, there are a number of passes along the route. These are the passes crossing the Swart, Kaaimans and Silver, Touw, Hoogekraal and Karatara Rivers, the Homtini gorge (or Goukamma River), and the Phantom Pass dropping down to the Knysna River.

Before this road was built, land access between George and Knysna was by no means easy. The first settlers in Knysna – the farm Melkhoutkraal, which included the whole basin containing the Knysna lagoon, was granted in 1770 – were probably attracted there by the beauty of the place. But, as Bulpin says, 'almost impossible lines of communication made the area seem more difficult to get out of than into, and only a South African bullock wagon made of the hard timber of the Knysna forests could have survived the first dreadful track that was blazed to connect Knysna to George'.

Be that as it may, there were those who went with their wagons between George and Knysna, although only very determined people would have tackled such a trip. George Rex, the almost legendary early landlord of Knysna, was one. He bought the farm Melkhoutkraal in 1804, and arrived to take occupation travelling with his family in a coach bearing a coat

The 1952 National Road between the Lakes and the sea (Kelvin Saunders)

of arms and drawn by six horses, and accompanied by a retinue of friends on horseback. However, the coach never left Knysna again – it was really not suited to the primitive tracks of that time and place.[1]

The first obstacle on the coastal route was the horrifying Kaaimansgat gorge, which succeeded in stopping east-bound wagons until Governor Van Plettenberg crossed, west-bound (the easy way!), with his wagons in 1778, after which this pass gradually came into use.[2] Another great obstacle was the dense forest vegetation encountered along the coastal strip. Thunberg, who came this way in 1772, recorded: 'The woods we passed through were dense and full of prickly bushes. We could find no other passage through them than the tracks of Hottentots, so that we were obliged almost to creep on all fours, and lead our horses by the bridle.'

From the early 1850s Bishop Gray and his intrepid wife Sophy travelled widely to attend to his pastoral duties. They recorded that there were seven 'dangerous' drifts between George and Knysna, and that the route was on occasions impassable to vehicles. They themselves rode on horseback, sometimes swimming with water over the saddle.[3]

A map in the Cape Archives (probably drawn about 1855) shows 'the road to Knysna' as crossing at Kaaimansgat and then keeping to the plateau until it turns sharply to the south between the Swart and Karatara Rivers. It then runs from north of Swartvlei, inland of Groenvlei more or less along the line of the present National Road, until it turns up the west bank of the Knysna River, presumably to the ford there.[4] I shudder to think of the topographic difficulties encountered. And the forests would have been a great hindrance.

Regarding the ford on the Knysna River, this was also restrictive: it was necessary to wait for low tide to cross, and even then the water came up to a horse's belly or to a wagon bed. As we have noted, Andrew Geddes Bain, when wanting to describe just how horrible a certain pass was, said: 'the fearful ruggedness outstrips even that between George and Knysna, and that is saying enough …'[5]

In short, there was no good overland route between George and Knysna before the Passes Road was constructed.

◆◆◆◆◆◆◆◆◆◆

Thomas Bain moved to Knysna in 1860 to build the streets and roads in that neighbourhood. In 1862 his works included the construction of a low-standard road to provide access to work the Zuurvlakte Crown Forest on the western side of the tidal Knysna River. He constructed the eight kilometres in two months, with a labour force of 22 convicts. It was not much of a road, built on a foundation of river boulders, but it was after all intended merely as an access road for foresters.

The road up from the river became known as the Phantom Pass, named, it is said, after the white Phantom Moth found in that area. It might be held that this pass, and the complementary stone causeway across the river at its foot, were the first bits of the Passes Road to George – even though it had to be upgraded in 1882 when required to carry through-traffic.[6]

From 1863 Thomas Bain and his brother-in-law, Adam de Smidt, who was the local District Engineer at that time, battled through forests and across river gorges, locating a

Timber bridge over the Knysna River built in 1893 to replace the causeway

route between George and Knysna. This was not all done at once: the location work along the whole route was spread over a number of years, as opportunity offered.

Apparently Bain and De Smidt had a strong difference of opinion over route selection, De Smidt holding that the line should be closer to the Outeniqua foothills so as to cross the rivers where their gorges were shallower. It is not unheard of for even latter-day engineers to differ as to the optimum location for a road and, if persuasion fails to convince the other chap that your selection is to be preferred, seniority prevails – and it is no use fighting against thunder. But the difference of opinion between the brothers-in-law was more serious, and it is in fact related that they never spoke to one another again.

The discovery of gold at Millwood in the late 1870s had a major influence on the final line at the eastern end of the project, resulting in the alignment beyond Lancewood swinging inland to serve the gold field instead of following the route of the existing (1860) track and the present National Road.[7]

◆◆◆◆◆◆◆◆◆◆

In 1861 a Select Committee was appointed to look into the possibility of connecting George with Humansdorp along the coastal plain. Not only would this avoid through-traffic having to scale the Outeniquas and travel down the Langkloof, but it would open up the coastal plain itself to development.

In 1867 the project in which we are currently interested got the green light, and Adam de Smidt commenced construction at the George end, working towards Knysna. It was to prove a difficult, long and wearisome job: although substantially completed in 1882, fifteen years later, the road was only handed over to the Divisional Council in 1887.[8]

The Kaaimans River crossing (Cape Archives J2842)

The first pass was down to the Swart River and up again on the other side. There was an existing crossing of sorts, known as Zwart River Hoogte on the old road, and on the far side this old road turned south towards the dreaded Kaaimansgat crossing. It was rather typical of the unplanned tracks of that time. Bulpin gives a good description: 'In former times the descent was a nightmare. The first wagon track simply slithered almost straight into the valley and with their wheels remmed (braked with blocks) the wagons wore the road into a channel more than 2 m deep and so narrow that a man could not pass between the banks and the side of the vehicle. Down the chuite the wagon went with a loud cracking of whips by the drivers to warn any travellers not to start coming up in the other direction, for a first-class disaster would occur if two wagons encountered each other on the pass.'[9]

To build a decent road through this pass, obviously on a new alignment, took considerable side-cut, with blasting the order of the day. The result was, as usual with Thomas Bain's and Adam de Smidt's works, an engineered roadway of an entirely acceptable standard.

It will be appreciated that the removal of the trees which stand in the way of any road can be a major factor in determining the speed of construction. As we have seen, the area in which we are interested was covered with dense indigenous growth when De Smidt moved onto site. The Great Fire, which opened up the coastal strip to road-making, went through this area in 1869, two years after commencement of the roadwork. Many of the forest giants would have been damaged in the fire, making their removal to clear the road reserve easier. But according to Burman, 'the hilltops were burnt clear, and only pockets of indigenous trees remained, sheltering in the deep kloofs'[10] – as De Smidt found, when he came to open the line for the passes down and across the rivers and up the other side.

By 1869 De Smidt's construction had reached the Kaaimans River and adjacent Silver

The present-day Silver River Bridge (Alf Balmer)

River gorges but work here was only finally completed in 1875.[11] By then a detached section of road between the Silver and Touw Rivers, and a four-metre construction road down one side of the Touw River gorge had also been built.[12] There had been a settlement here (nowadays called Ginnesville, where White's Road to Wilderness takes off, 14 kilometres from George) since the 1750s,[13] and the access road to the settlement enabled the *padmakers* to open up a separate construction site here.

In fact, because of the nature of the terrain and difficulties of access, the whole of the construction on the 75 kilometres was very fragmented, with bits being done here and there as was most convenient and economical. Thus we find that Bain's annual reports reflect this phenomenon: it was not a case of 'colouring-in' progress as an ever-lengthening worm on the plan – progress more often looked like a broken necklace.

The main camp was sited where Saasveld stands today, between the Swart and Kaaimans Rivers. A sub-camp to house 50 convicts was also opened near Ginnesville to work that section, and subsequently other camps were established as construction progressed towards Knysna.

Thomas Bain, for whom the Passes Road was only one of his many responsibilities, was transferred with his family from George to Tradouw Pass in 1869 and to Hermon (to build a railway line through Nieuwekloof) in 1873. One must assume that the experienced Adam de Smidt had less 'hands-on' advice from Bain during this period, although records show that Thomas travelled prodigiously to keep in touch with his various responsibilities.

The next river pass to be tackled was at the Touw River. Previously known appropriately as Trek-aan-Touw, a corruption of the Khoi 'Krakede-Kau', meaning 'maiden's ford', this river had to be crossed by those travelling the old road between Kaaimansgat and Duiwelskop Pass, apparently straight down one side and straight up the other with every ox and person 'pulling on the rope'. Victorin in 1854 described the Touw River approaches thus: 'situated deep down, so the road down to it [he was travelling towards George] is extremely difficult. The oxen were exhausted, so that when we were well down at the bottom of the ravine, 12 pairs of oxen were inspanned to the one wagon to pull it up to the top of the slope … Another small river, tumbling into the large one just below the drift, forms a waterfall 10–15 alnar [6–9 metres] high between the perpendicular cliff masses more than a hundred alnar [60 metres] high.'[14] Bain and De Smidt, with their construction facilities, were able to build a road angling across the contours at a steady and acceptable gradient on both approaches to the river.

Up the eastern slope of this pass, 20 kilometres from George, is the second link to the Lakes, via Hoekwil. Six kilometres further on is the forestry road leading northwards to the old Duiwelskop Pass. Construction reached this intersection in 1871.[15]

The third connection to the Lakes, at Rondevlei, is passed before the road crosses the Diep River, also known as the Klein Keur and sometimes as one of the many Swart Rivers in this region, where the run-off is stained dark brown by rotting vegetation. It is more of a river crossing than a pass.

Thirty-four kilometres from George, the hamlet of Lancewood is reached. Initially the construction swung southeastward at this point, towards the sandy area at Ruigtevlei – there is today a road along this line – to join the then-existing track to Knysna, which ran, as mentioned earlier, very roughly along the present National Road alignment. At that time the pace of construction seems to have slowed dramatically, possibly because the authorities were

The Trek-aan-Touw approach in 1816 (C.J. Latrobe)

concentrating their major efforts on improving routes to the Kimberley diamond fields and the burgeoning gold fields in the Transvaal.

The first gold nugget was found in the Karatara River bed in 1875, but it was only really after an *in situ* meeting in February 1879 that the Ruigtevlei line was summarily abandoned and Bain's efforts were redirected to a new route, springing off from the original line at Lancewood and heading initially eastward.[16] This higher route, which it is said had been favoured by Bain all along, would serve the new gold field on its way to Knysna. But it had to cross some river gorges to do so.

Thus the road ducks down the Hoogekraal (a small pass) and Karatara River valleys. The fifth and sixth links to the Lakes – they join the previous road on its way to Ruigtevlei – take off, one on each side of the Karatara River. (These link roads are mentioned to assist the earnest seeker in determining his position should he wish to inspect the Passes Road.)

Drift in the Homtini Valley (Cape Archives J9360)

The route then turns southeastward towards Knysna and the crossing of the dreaded Homtini gorge. This pass was a major construction feat, with the road curling down two and a half kilometres to the bottom of the gorge and up the other side, through dense indigenous forest. Bulpin describes the pass as 'a classic piece of old-time roadmaking with dramatic views and the indefinable elegance of its curves. The river itself is a gorgeous torrent of amber water, tumbling down from the deep forests of the mountain slopes to the north.'[17] This most attractive pass was in effect completed in 1882, but was only available for traffic the following year.

Construction continued, to join the Phantom Pass on the western slopes above the Knysna River. Bain's 1862 foresters' road was renovated to take the main road traffic now directed along it 20 years later.[18] However, the basic concept of the design needed revision and we find that in 1889 Thomas Bain had surveyed a new route and that in October of that year construction started on the new pass.[19]

The Passes Road was completed in 1883, sixteen years after work commenced on it. It was the main road between George and Knysna for almost 70 years, until the National Road was completed in 1952, and still serves the area through which it passes.

CHAPTER TWENTY-ONE

SWARTBERG PASS

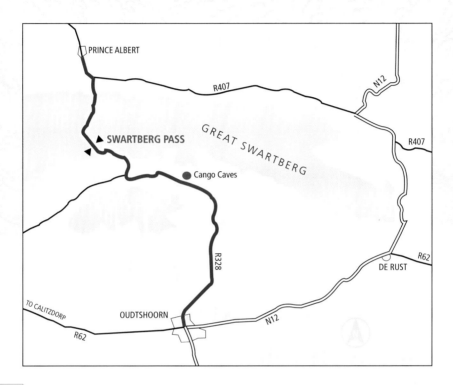

The Swartberg range was in the early days a nearly insurmountable barrier between the Little and Great Karoos. The inhabitants of the Great Karoo were consequently ill served with transportation links to possible markets because of the unfriendly terrain which intervened, and very much wanted to take advantage of the coastal shipping which called at Mossel Bay and served the folk in the Little Karoo so beneficially.

Meiringspoort and Seweweekspoort were the obvious routes through the Swartberg. Roads were constructed through these two poorts in 1858 and 1862 respectively, and although necessarily of a low standard because of limited funding, they gave a considerable economic boost to the Great Karoo.

Unfortunately the dire forecasts of the engineers (who had asked for more funds to build to a higher standard) were proven to be correct: these roads, crossing and recrossing their respective rivers through the gorges, proved unreliable. They were frequently damaged by floods and subsequently closed to traffic, sometimes for months on end. But the locals had 'tasted' the benefits of a transportation link to Mossel Bay, and now wanted a more reliable crossing of the Swartberg. So they petitioned for a direct road link between Oudtshoorn and Prince Albert in 1878 and again in 1879.[1]

Swartberg Pass (Cape Archives R1680 and AG1053)

Thomas Bain, who had suggested the construction of Swartberg Pass as a possible solution to the flooding problems in the poorts, got the go-ahead to investigate the feasibility of his proposal in 1878.

Although a start had been made to open up a bridle path from the Prince Albert side, and indeed £600 had been spent on the initial section, there was no existing defined route over the mountain. After considerable walking and riding Bain found four possible lines, from which he selected and planned the optimum location. The maximum gradient was 1 in 8, compared with the 1 in 6 of Montagu Pass. His July 1879 estimate of cost for the 24 kilometres was £10,418 by convict labour (which was free) and £24,942 by 'free' labour (which had to be paid for). The pass, besides having the advantage of providing an all-weather road, reduced the distance from Oudtshoorn to Prince Albert by 54 kilometres as compared with the alternative through Meiringspoort, which was of course quite a consideration for transport by ox wagon.

The enthusiastic Prince Albert Divisional Council had made a good wagon road to the foot of the gorge on the northern side by July 1879, and by December of that year they had established a bridle path to their boundary. However, it was July 1880 before parliament finally approved the report, plan, section and estimates for the pass, and things began to move.[2]

The work was put out to tender under the 'new system' of James Fforde, the former Chief Inspector of Public Works. This system was to test whether passes or roads could be built more economically by tender than by convict labour. (Actually, records show that this was not the first contract for pass-building. For example, in 1819 S.J. Cats contracted to build the

Towering cliffs cradle the northern entrance to Swartberg Pass (P. Wagner, courtesy of Photo Access)

The hairpin bends on the way to the valley below (Peter Steyn, courtesy of Photo Access)

Franschhoek Pass, and in 1862 the Oudtshoorn Divisional Council accepted the tender of D. Hattingh for £2,000 for the construction of Schoeman's Poort, just around the corner from the Swartberg Pass.) At any rate, tenders were invited and eight were received. The highest was for £63,784; the lowest, from Jan Tassie, for £18,120 with an 18-month construction period was accepted.

Tassie initially imported 101 labourers from Delagoa Bay, but their numbers steadily decreased owing to desertions and he had difficulty in recruiting locally. He started work in October 1881, completing only five and a half kilometres and receiving only £4,098 before he was declared insolvent in January 1883. There was the usual delay while the contract termination was sorted out, and before a decision was taken to complete the pass departmentally, using convict labour.[3]

By November 1883 Thomas Bain was able to start construction, with 200 to 240 convicts, the man in charge on the site being John Fitz-Neville. By the end of 1884 the construction from Prince Albert had reached the summit, and one and a half kilometres of bridle path, five feet wide, had been built on the Oudtshoorn side.

Work was set back in May 1885 when heavy floods washed away sections of the newly constructed road. But the lesson was well learned, and Bain made sure that the new construction was above the flood levels. In hindsight (the only precise science), at a time when topographic and rainfall data were not available to calculate run-off, he was probably lucky to have a good rain at an early stage to indicate to him the flood levels along the works – although I doubt whether he felt that way at the time.

He was able to open the pass for light traffic from March 1886, and from September it was opened to carts daily and to wagons on Fridays, and a regular post-coach service was being operated. The pass was officially opened on 10 January 1888 by the Commissioner of Crown Lands, Colonel F. Schermbrucker, with the requisite champagne, 21-gun salute and so on. His statement that 'ten thousand travellers will in future feast their eyes on this beautiful picture' has without doubt proved to be correct. And I am sure that the zigzag ascent from Muller's Kloof on the northern side, and the magnificent dry-stone retaining walls which Thomas Bain built, have been photographed almost as often as the scenic views.

The final cost of the pass, including two miles of approach roads at either end, was £14,500, not counting convict labour worth £17,000. There can be no doubt that poor Jan Tassie with his tender of £18,120 sadly underestimated the costs involved, or that A. Smith of Sea Point, who submitted the high tender of £63,784, should have been able to make a profit had he been awarded the contract.

One can only agree with William Grier, the Chief Inspector of Roads at that time, when he said: 'It is a work of bold conception and has been a heavy and difficult task to carry out, and one which reflects great credit on Mr Bain for the ingenious manner in which he overcame the physical difficulties he met with.'[4]

And so the inhabitants of Prince Albert, and of the Great Karoo in that vicinity, got their all-weather connection via Oudtshoorn to the port at Mossel Bay. Well, it was pretty safe from all but (comparatively) minor damage from floods and washaways, but nothing in this world is perfect and the pass, rising to an altitude of 1,583 metres above sea level, has throughout its lifetime been closed for an average of one or two days each year because of a problem not experienced in the poorts: snow.

The road Thomas Bain constructed is basically unaltered to this day, and the surface is still gravel. You will find that the locals are against the pass being improved and surfaced. Their reason? The extremely low accident figures on this route with its low geometric standards.

This was the last of the many passes built or supervised by Thomas Bain in the Cape. On 10 February 1988 a memorial plaque was unveiled near the summit by the Administrator, Gene Louw, to mark the 100th anniversary of the opening of Swartberg Pass. The pass was declared a National Monument later in the same year.[5]

Near the foot of the pass is a road sign which says that one may not take a caravan over the pass. That sign is old, rusted and shot full of holes but, speaking as one who has towed caravans through a number of places where a caravan had never been seen before, I suggest that the inscription is pretty good advice. You should unhitch and drive over the pass with your tow vehicle alone. But allow plenty of time, especially if you are a *padmaker*, because you will want to stop, climb out and look more closely at a number of Thomas Bain's construction details – as well as wondering at the magnificent views and listening to the silence.

CHAPTER TWENTY-TWO

PENHOEK PASS

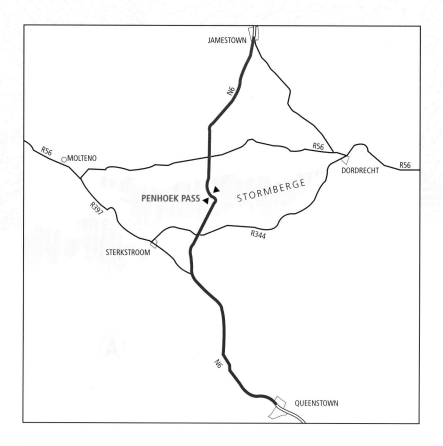

P enhoek Pass lies on National Route 6, about 60 kilometres north of Queenstown on the way to Jamestown and Aliwal North. It is a much 'younger' pass than those which have been covered in preceding chapters, as pass-building spread out from Table Bay and later from Port Elizabeth, with not much call for this sort of activity in earlier days north of Queenstown.

The first mention of Penhoek Pass which I have come across (there are undoubtedly others) is in the Cape of Good Hope Roads and Passes Maintenance Act of 1892. Schedule A lists only 17 passes which are entitled to contributions from the colonial government. Penhoek is down for £400 for 'repair and maintenance' in 1892/93, and a subsequent annual maintenance contribution of £200. So there must have been something there at that time.

A map appearing in *The Times Atlas* of 1895 does not show a road crossing the Stormberg range between Queenstown and Jamestown: the road and telegraph route went via

Dordrecht. A map illustrating aspects of the Stormberg Campaign during the Anglo-Boer War, published in 1900, also does not show a direct road. In fact, this map does not even show a road between Dordrecht and Jamestown. The Queenstown Automobile Club's *Motorists' Route Book*, dated 1924, does show a road over Penhoek Pass, but with the annotation 'Note: the direct road from the bottom of Penhoek has been abandoned in favour of the route via Halseton' and presumably via Dordrecht.[1] A further annotation says the road is 'out of repair but promised ultimate renewal'. All was obviously not lost.

The National Road Board included the Queenstown–Aliwal North section in their proposed network for the original proposals in 1936. However, it is interesting to note their ambivalence regarding Penhoek Pass: in 1944 Route 6 is described as Queenstown–Penhoek (or Dordrecht)–Jamestown–Aliwal North.[2]

Hennie Aucamp gives an interesting description of the early pass:

> According to a land surveyor, the late S.L. Moorcroft, who owned the farm across which the road passes, the name dates from the days of the ox wagon. The original road followed a kloof or narrow glen in the mountain, the top part crossing a broad rocky face several metres below the last ridge. The rock was so slippery and the gradient so steep that the oxen were unable to pull a wagon up. A peg was therefore fixed in the rock ledge, and as soon as the oxen in front reached it, the whole team was outspanned and a long chain was fastened to the wagon shaft. The chain was twisted once around the peg and then hooked again to the team, which was now facing downhill. In this way the oxen found it easy to pull the wagon uphill while they were straining downhill. And once this steepest and most difficult section was passed, the last stage was easily covered. When the present road was built in 1952, the rock face was unfortunately blasted away and the peg disappeared.[3]

Fresh ox teams were apparently available at the farm Buffelsfontein at the crossing of the Molteno–Dordrecht road. Incidentally, as at 1992 the coldest temperature ever measured in the Republic was recorded on this farm.

The signboard at the summit of Penhoek Pass once showed 6,051 feet (or 1,844 metres). Danie Ackermann, the Resident Engineer here in the 1950s, says that there were sarcastic comments about the final '1', but as he rightly says, this is what it measured at, so why not show it on the board? My point is that this was a very high pass. Besides the very high Naudesnek–Pot River Pass (2,492 metres), there are few passes which reach the same altitude as Penhoek.

Pieter Baartman, who became Province's Chief Surveyor, has provided his reminiscences of two contacts with the pass.

'My earliest experience of Penhoek was in January 1940 when I was eleven years old. It was a proverbial dark and stormy night. Along with my sister I had been on holiday in the Free State, travelling there by train but returning in this manner only as far as Aliwal North. My father, with a co-driver, had arranged to meet us there and take us home to Queenstown by car.

'We arrived at Penhoek and drove into a mist so thick that even the powerful headlights of the Riley sedan could not penetrate it. The pass was still in its original state, rough gravel surface, narrow, winding and, although not all that long, on this night it seemed to have no end. My father walked ahead of the car, just visible in the light of the headlamps, scouting out the line of the road with a torch that was practically useless against the fog. So it was almost a case of feeling one's way step by step. A journey which should have taken no more than a quarter of an hour progressed at a very slow walking pace with occasional stops for the co-driver to do a personal check on just where he was going!

'Within a month of that night my elder brother Willem, newly matriculated at fifteen years of age and having joined the Roads Department as a Survey Assistant, arrived at Penhoek having been given the task of staking out the line for new construction ... When I started service it was almost a year before I was trusted to use a tacheometer unsupervised; here was Willem actually pegging out the curves of a pass within a month ...

Penhoek Pass under snow (Danie Ackermann)

'By the start of 1941, with many of the Department's men in the Engineering Corps in North Africa ... several of the construction units had closed for the duration of the war. The remaining units were brought up to strength by transferrals from those that had closed, and this brought Mr J.A.C. Bester to Sterkstroom as Senior Foreman under the Resident Engineer. Mr Bester's specific task was supervising the construction of Penhoek Pass. He remained in Sterkstroom from early January till late in the year, when this unit also closed and moved to Stutterheim. The pass construction had by then progressed to the point where the easier grade of the new line met and crossed the steep track that was the old road. The final feature built was the only box cut on the pass ...

'I transferred to Jamestown in 1950 with instructions to supervise the establishment of the unit and to restake the whole of Route 6 Sections 4 and 5 from Bailey through Jamestown to Aliwal North. The road between Queenstown and Bailey was already tarred (low standard) and from Bailey to Penhoek the existing gravel road, though well built on reasonable vertical alignment, needed to be upgraded to meet the latest standards ...

'I could never get enough of the magnificent view from the summit of Penhoek to the south and the hills around Queenstown. During my four years at Jamestown I experienced the vagaries of the Stormberg weather on many occasions ... I found myself staking out an up-graded route in Penhoek, with snow up to my ankles and deeper, my fingers adhering to the metal clamps of the instrument ... In fair weather Penhoek was a pleasant place to be, but summer was the season for thunderstorms of quite alarming proportions and winter brought the dense mists and snowfalls that made it a dangerous place.'

Danie Ackermann was at Jamestown more or less over the same period as Pieter. Danie also comments on the cold weather they experienced.

The new construction and the old road (Danie Ackermann)

'When I assumed duty as Resident Engineer in Jamestown in June 1951 the construction camp, situated on the commonage, was still in the course of being erected. I awoke the first morning to find snow falling and the countryside covered in white. It took some considerable time to appreciate and come to grips with the winter climate in the Jamestown area and Penhoek Pass in particular. In Jamestown itself I personally measured a minimum temperature of minus eight degrees Centigrade.

'At that time the practice was to drain the radiators of all construction plant overnight in winter and to assign labour to start heating water in 44-gallon drums at about 4 a.m. for refilling at 7 a.m. when work was to commence. Unfortunately the water froze almost as fast as it was being poured into the engine. Truck batteries were often run flat attempting to start the vehicle, but once one was started the others would be tow-started. Then it was the turn of the tractors and power shovels, which could not be towed. After switching batteries back and forth, the tractors and power shovels would be started. Often construction could only commence about 10 a.m. After representations to Head Office, permission was granted to purchase special starting capsules 'on an experimental basis' – which later became the rule.

'Accommodation in the camp consisted of prefabricated houses for the senior staff and so-called "pied-à-terre" houses for the other staff. The latter were a new type of house on construction units and consisted of walls made of a soil-cement mix compacted between sliding shuttering. Given the climatic conditions these proved very comfortable. In contrast, the prefabricated houses consisted of a steel-framed building clad with inner and outer skins of asbestos-cement sheets. For insulation the wall cavity should have been filled with vermiculite, and the ceiling insulated also. For cost reasons this was not permitted. I remember sitting through a full morning's discussion at an engineers' conference on steps that could be taken to keep these houses cool in summer. At the close of that discussion I asked for consideration of measures to keep the houses warm. This request was not considered worthy of discussion.

'Temperature inside these prefabs was the same as outside – in other words, well below freezing in winter. The official ration of fuel for cooking and heating was one wheelbarrow each of coal and wood per week per household, but most of the heat provided by Queen stoves was absorbed by drying nappies or disappeared through the gap between wall and ceiling. Even on clear sunny days nappies, hung outside, would freeze as hard as planks ...

'Special plant for snow clearing of the road was of course not available. A road grader worked satisfactorily. A single lane to the bottom of the pass had to be cleared before turning round to clear the other half of the road and to assist vehicles that had gone off the road through either skidding or not being able to distinguish the roadway. Somehow the occupants survived the night in the frigid conditions – these were the days before car heaters were common.'[4]

These two accounts, I feel, give a good insight into the lives and doings of the Cape Provincial *padmakers* in the mid-1900s. Pieter Baartman's contribution, because of his earlier contacts with Penhoek Pass, also gives a picture of the progress made over the years to improve traffic conditions there.

The pass was reconstructed to improved standards in the 1980s. It is a pleasure to tow a caravan over it today – the original 'pen' is definitely no longer needed. Don't forget to stop at the summit to enjoy the wonderful view.

CHAPTER TWENTY-THREE

HUIS RIVER PASS

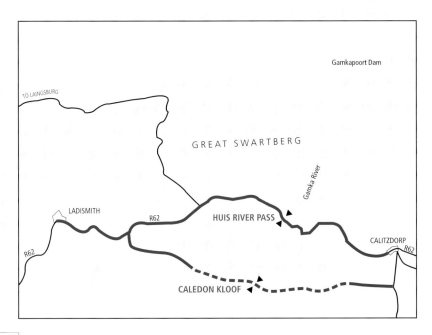

Huis River Pass is on the Ladismith–Calitzdorp–Oudtshoorn road, not far to the west of Calitzdorp. The pass provides a connection between the Ladismith and the Oudtshoorn valleys of the Little Karoo. It also makes possible a much-needed additional east–west link in the road network, lying as it does between the N1 (Laingsburg–Beaufort West) link to the north of the Swartberg, and the N2 (Riversdale–George) link south of the Langeberg–Outeniqua mountain chain.

It is not an old pass: it was first opened as a constructed pass in 1896. Prior to that, the (grudgingly) accepted route was via Caledon Kloof. This was truly a horrible pass.

CALEDON KLOOF

Caledon Kloof was discovered in about 1807. Its lies to the south of the Huis River Pass, in the poort through the mountains made by a stream. Goetze says that it was opened up by Gerrit Pretorius and, being the only direct way to Ladismith, was used extensively.[1] It is recorded that in 1866 there were potholes half a metre deep alternating with huge rocks, and that the kloof was – not surprisingly – 'littered with broken wheels, bits of wagon and skeletons of oxen'. The kloof is narrow: too narrow to carry both the road and a rail line, as confirmed by this extract from the report of a 'flying survey' or reconnaissance to investigate this possibility, carried out in 1882. 'This kloof is very narrow, and has perpendicular krantzes of hard sandstone on both sides; the whole breadth of the valley being occupied by the river and road in some places. The road must be sacrificed for the railway, and the former would have to be taken by way of Huis River.'[2]

The kloof was originally called Welgevonden or Rooielsboskloof, and some knew it as Verkeerdekloof.[3] These differences do not matter as everyone is agreed that it was renamed Caledon Kloof (after the Governor) in 1810, possibly in the vain hopes of getting some funding to build a road there. (For all that, it is marked Verkeerdekloof on the topo map.)

The whole bed of the kloof was at any rate totally washed away by a flood in May 1885, when widespread flooding did considerable damage to many roads in the Cape, including washing away every trace of the constructed road through Meiringspoort.[4] And that, as far as I know, was the end of any thoughts of developing Caledon Kloof as a permanent route between the two valleys.

HUIS RIVER PASS

Huis River rises somewhere north of the Klein Swartberg, flows through Seweweekspoort, and sidles past Huis River Pass to join the Gamka River, whose combined waters then join those of the Olifants to flow under the bungee jumpers on the old N2 bridge as the Gourits. I had occasionally wondered idly as to which 'house' the name referred, and so was interested when I found that no house was involved: the name is derived from the Khoikhoi word for 'willow tree'.[5]

There must have been some sort of track through the mountains for the authors of the railway flying-survey report to have so high-handedly shouldered the road out of Caledon Kloof in 1882. Indeed there are wheel grooves visible in the rocks to the north of the present Huis River Pass where a bypass of sorts was made for part of the 1964 construction.[6] However, the line of this track was not even considered when the Calitzdorp and Ladismith Divisional Councils combined forces to build a road through the pass in 1896–7.[7] As we have seen, the line which wagons must take up a steep mountainside is quite different from the line which the engineer, with the benefit of a flat road-bed to prevent wagons from toppling over sideways, can select to reduce the steepness (gradient) of the road.

This original Divisional Council pass was an extremely good piece of work, and served well for many years. True, bits of the mountain fell off and blocked it now and then, but I am sure that any complainants were handed a shovel and reminded of the alternative: the horrible Caledon Kloof. In the 1950s the Divisional Council reconstructed the roadway, and this improvement kept things going for a further number of years.

But then came the great age of the Trunk Road scheme, constructing all-weather roads to communities where access had previously depended on weather conditions, and enabling the local road authorities – the Divisional Councils – to permanently surface many of these roads, thanks to generous and far-sighted Provincial subsidies. It was a great era of road-building in the Cape, and I was privileged to have been involved in parts of it. It had unfortunately to be curtailed for financial reasons, like so many other good things. As part of this scheme the road through Barrydale, Ladismith and Oudtshoorn was constructed to then-current standards, and permanently surfaced. This entailed improving Huis River Pass.

The Divisional Council construction, 1930 (NLSA 19465)

The really critical pass section is that which climbs up a valley where the mountains to the north push against the mountains to the south, leaving very little room indeed for a road. The southern mountains are too indented and broken to encourage anyone to think of putting a road there, while the northern mountains have been inclined by folding in such a way that any bits of rock which break off or weather away find it natural to slide downhill onto the road. To be technical for a moment, the side slope has a ratio of approximately 0.8 vertical to 1 horizontal, which is very close to an average safe-rest angle for the surface. The section we are interested in is underlain by fine-grained argillaceous sediment of the Kango formation that has been metamorphosed under severe thermal and dynamo-metamorphism into phyllites possessing high shear characteristics and a series of well-developed joints, resulting in many potential slide planes often closely paralleling the existing natural surface. Thus Godfrey Floyd, geologist.[8] What this means is that if you dig into the toe of the mountain slope to widen the road, you are liable to have things sliding down onto you and the road.

| *The successful semi-retaining walls* | *The 1966 construction (Graham Ross)* |

I was Acting District Roads Engineer there in 1951, and I was asked to see if I could find an alternative route that would not climb up a valley with such unfavourable geology. John Williamson and I studied the topographical maps (there were no aerial photos available at that time) and agreed that there was no point in looking for a route through the broken country to the north. There was also very little hope to the south, but at least it did not look as repelling as the northern side.

So, armed with what maps were available, my binoculars and an Abney level, I spent days driving and climbing onto the mountains to the south, accompanied in relays by teams of very helpful locals recruited by the Calitzdorp Divisional Council. My main memories are of intensely sore legs. For the six months prior to my sudden transfer to Oudtshoorn I had spent my working hours sitting in the bridge design office and my weekends sailing. Neither the design office nor a seven-metre cutter gave much practice in walking, so my legs were not in very good shape. I still remember with pride the great surprise on the faces of the farmers when I turned out, hobbling but game, for the second day of our expedition. Of course, I was younger then.[9]

We did not find a feasible alternative route, and the new pass had to be built up that frightening valley. What with one thing and another, more than ten years passed before a 'best' design had been decided upon, and a contractor (A.G. Burton) was on site.

Kantey & Templer's Resident Engineer, 'Army' Armstrong, and Burton's Site Agent,

The 1966 construction (Graham Ross)

Philip Wessels, lived in Ladismith – Army had a four-gabled ostrich feather-boom house with eleven rooms – while the rest of the contractor's staff lived in prefabs in Calitzdorp. They were cut off from their homes by flash floods on two occasions, once when the Gamka River topped the old bridge, and once when a culvert on the old road was washed away during a storm.

Because the rock was so steeply bedded, in places even a small blast caused large slides which blocked the old road and the formation of the new construction. As a result, although not originally intended, a bypass was constructed through some unamenable country to the north, around this area, to ensure that the pass could be kept open for normal traffic.

The drilling was carried out using wagon drills, which were winched down the slopes on steel cables. On some sections the slopes were too steep for the blasted rock to be moved with trucks, so two crawler tractors with drawn scrapers were used to drag the initial fill into place.[10]

Hydraulic excavation was used to stabilise slopes where needed, and especially where the toe had been cut away. High-pressure water jets, directed into crevices between the rocks, saturated the thin clay layers, lubricating them, and thus inducing a slip. The resulting debris on the formation below was dozed away – when it was reasonably certain that no more rock intended falling down – to form part of the new road. I was fortunate enough to be on site when the South African equivalent of the lumberjack's 'Timber!' drew my attention to the fact that what looked to me like half the mountainside was sliding down towards the construction shelf where I was standing. It was a most impressive sight. Army then told me that this hydraulic lubrication was a dangerous business for anyone on the lower levels where the rocks were likely to end up – which I could by then see for myself. Apparently on a previous occasion Army and Philip were observing the jetting when, out of the blue, a big fall of rocks came hurtling down. A rock as large as a small car landed five feet from Army, while a smaller boulder hit Philip on his calf muscle and he could not walk for a few days. Very shortly after this I had urgent business which took me elsewhere!

As a protection for the travelling public – and the road – Kantey designed a series of heavily reinforced 'semi retaining walls'. These were placed along particularly unstable sections, especially where the toe had been cut away. They were erected in such a way as to leave a cavity behind them, providing holding areas to catch falling material, including rocks. This could be cleared away by the normal maintenance crews at a time suitable to them, to renew the availability of the cavities for their designed job. The unique scheme has worked very well indeed. The walls show occasional star-cracks where the heavy reinforcing has enabled them to withstand the punch of some particularly large boulder, and the wall tops are a trifle frazzled in a few places, but the detritus caught behind them, waiting to be cleared away, shows that they are fulfilling their designed purpose.

This gives an excellent idea of the steep side-slope (Graham Ross)

When travelling over the pass, one's attention cannot but be attracted to the neat cut slopes in the 60-foot-deep box cut at the top of the pass section proper, and at a couple of other places. This was attained with pre-split blasting where the rock was suitable, which entailed drilling holes down to road level three feet apart, sloped to the design profile (plus holes between the edges, three feet apart in both directions, naturally). These were loaded with charges three feet apart vertically, with sand between them. The whole set-up was then exploded throughout the prism. There was no big bang, just a shudder, and it worked like a charm. The dozers cleaned out the material in the prism, down to road level, with little effort.[11] The half-drill holes along the sides of the cuts add to the neat appearance.

The reconstructed and surfaced pass was officially opened on 6 May 1966. It certainly is a most spectacular piece of roadwork, set in majestic surroundings. The *padmakers* have provided a couple of picnic places along the section so that one may pull off and get out to appreciate what has been achieved here.

CHAPTER TWENTY-FOUR

CHAPMAN'S PEAK DRIVE & THE MOUNTAIN ROADS TO HOUT BAY

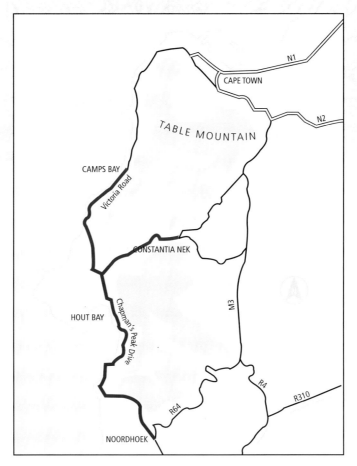

C hapman's Peak Drive, connecting Hout Bay and Noordhoek on the western coastline of the Cape Peninsula, is an outstanding construction feat, and a unique and impressive scenic drive. Chiselled into the nearly sheer rocky mountainside, it has been accepted from the time that the project was first thought of that continual minor maintenance would be needed to clear the roadway of rockfalls.

Unfortunately, through the years strata in the exposed cliffs above the road have gradually weathered and softened, leading to an increase in the frequency of rock falls. Following a tragic incident in January 2000 when by a one-in-a-million chance a rock fell on a car, fatally injuring one of the occupants, the road was closed to traffic while various courses of action were being considered.

The Hout Bay valley has two other access roads, one over Constantia Nek and the Victoria Road from Camps Bay. All these routes have very interesting histories.

THE CONSTANTIA NEK ROAD

Originally known as Cloof Pass, this was the earliest road into the valley.

The early Dutch settlers initially obtained timber from the Kirstenbosch area, and their first road was built from the fort to this source. Then it was found that much excellent timber was available in the Hout Bay valley (hence the name, of course). Commander Wagenaer decreed that an access should be opened up over Bosheuwel to this forest area, and on 18 August 1666 Lieutenant Schut and 24 soldiers went out 'towards the evening with some crowbars, picks and shovels in order to make a road'.

As demand grows, so roads have to be improved to meet the demand. So we find that Commander Simon van der Stel lengthened and improved the Hout Bay road in 1679, and that further work was done in early 1693.[1]

The colourful French traveller François le Vaillant, who has left us so many picturesque descriptions of places and events, described the road in 1785 as being the most beautiful in the surrounding countryside.[2]

By 1787 the road had reached as far as the farm Kronendal, but unfortunately after that the road was broken up, with deep excavations, as part of the French Regiment's protections against a possible invasion by the English. And, from the fact that William Duckitt in 1800 complained that the road was almost impassable for wagons, it looks as if any reinstatement was of a decidedly casual nature. However, undoubtedly as a result of complaints from travellers, a major effort was made to improve the road in 1804.[3] And so things went on, until in 1933 traffic had increased sufficiently for the road to be reconstructed and surfaced as far as the Hout Bay Hotel, and it has subsequently been further upgraded as demand increased.[4]

It is always difficult to decide where to cut one's account in describing road histories, but by a determined effort of will I have not included the rather fascinating history of the feeder road to Constantia Nek: De Waal Drive and Rhodes Drive. You will have to look elsewhere for that story.

VICTORIA ROAD

The second access route to Hout Bay is via the Atlantic coastline from Sea Point and Camps Pay, past Oudekraal and other picturesque little bays, over the neck above Llandudno, and down the slope to the village.

In 1848 a road was being built over Kloof Nek behind the city, and along the coastline a track was developed between Camps Bay and Hout Bay.[5] However, these were pretty basic roads. Storrar says 'they were nothing more than dusty tracks in summer, squelching quagmires in winter'.[6]

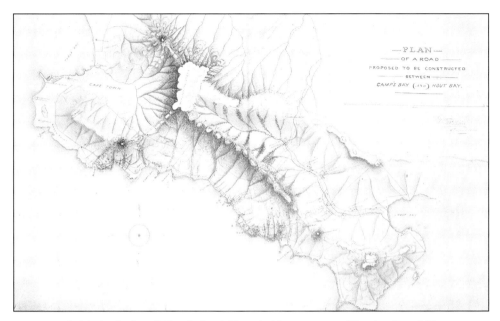

Bain's sketch plan of Victoria Road (Cape Archives, courtesy of Gunther Komnick)

The first real road between Sea Point and Hout Bay was constructed by Thomas Bain, only in 1888. It was a joint effort, as many of these major works had of necessity to be. The colonial government provided Bain and the necessary convict labour, and the Divisional Council put up £1,500 to meet some of the costs. Work commenced near the end of 1884, and two years later the section as far as Camps Bay was open over weekends for use by the public, while construction struggled on further along the coast. The road right through to Hout Bay was opened in a fashion to public traffic by mid-1887, but construction was not finalised until mid-1888.[7] (All road engineers have experienced this pressure to open a facility to the public while work is still going on. The resulting interference with the efforts of the dedicated *padmakers* makes one think of the benefits which would accrue to our country if the highest places were occupied by road engineers only.)

Subsequently the road has been improved, upgraded and surfaced at various times, but it still basically follows Bain's line and, besides being a magnificent scenic drive, has served Hout Bay and its inhabitants well to this day.

On 29 September 1993, one hundred years after Thomas Bain died, a plaque commemorating his work was erected near the highest point of Victoria Drive. And on 24 September 2000 a replica of his gravestone, as well as a suitably inscribed plaque, was also unveiled near there, in the parking area adjacent to the road above Llandudno.[8]

CHAPMAN'S PEAK DRIVE

In 1913 motor cars were becoming more common on the streets of Cape Town. In that year it was accepted as government policy that an 'All Round the Cape Peninsula Road' (ARCPR) suitable for motor cars should be planned and provided. Funds were made available by the Minister of Finance – some said as part compensation for removing the national capital to Pretoria – and the Minister of Railways provided 750 convicts at no cost. Construction started the same year on De Waal Drive, the Miller's Point–Smitswinkel Bay–Klaasjagerberg road and the link to Cape Point, and on the Red Hill road, at first by the Union Public Works Department, but later by the Cape Provincial Administration. In 1915 work began on Chapman's Peak Drive, which was finished in 1922. The final link in the ARCPR was completed when the Witsand–Slangkop section was opened in 1923.[9]

However, to return to Chapman's Peak Drive. It is said that the Magistrate of Cape Town, supported by a Mr Bromley, originated the idea of linking Hout Bay to Noordhoek. There was an exchange of ideas between the Cape Publicity Association and the Commissioner of Public Works on 25 January 1910. The PWD carried out 'surveys', put in a negative report, and the proposal was turned down.

In 1914 the Administrator, Sir Frederic de Waal, appointed the mining engineer Charl Marais to investigate the route. An important part of Marais's investigation was his study of the geology of the cliff. He recognised that the lower third of the mountain face was hard Cape Granite, the top of which was reasonably flat, and that a layer of soft, thinly bedded sedimentary rocks of the Graafwater Formation lay directly on top of the granite platform.

An aerial view of Chapman's Peak Drive (M.J. Mountain)

This 'soft' rock could be excavated to accommodate a road with the minimum use of explosives – there was a labour force of 700 convicts allocated to the project – and the horizontal layering of the sedimentary deposits made possible a stable, near-vertical face to the cuttings. Marais's recognition of these geological factors, showing that it was possible to cut the road into the softer layer along the face of the precipitous cliff, made all the difference to the feasibility of the daunting project.[10]

Sir Frederic, the first Administrator of the Cape Province, was obviously the main force behind the project, doing the things that Administrators are there to do. Following Charl Marais's report, Sir Frederic called in William C. West – apparently an esteemed mountaineer – to determine the route, and to flag it for construction.[11]

Work on the road started from the Hout Bay end in April 1915 and from the Noordhoek end in June of the following year. The Union government provided convicts at no charge (a continuation of the policy mentioned above), while further finance this time came from the Provincial Administration.

The engineer-in-charge was Robert Glenday, AMICE, who was to take over from W.L. Trollip as Provincial Chief Inspector of Roads (the post later renamed Provincial Roads Engineer) on completion of the job. He did the final surveying, designing and fixing of the line for construction, besides controlling the work of his technical and supervisory staff,

The horizontal layers of sedimentary rock are clearly visible (M.J. Mountain)

Convicts excavating by hand in the softer sedimentary rock (Hout Bay Museum)

and the 700 convicts who clung like flies to the mountainside while carrying out the hard manual labour of the construction.

By September 1919 the road was open from Hout Bay to the lookout at the highest point on the pass, while work on the really tough, near-vertical section further on continued. A word of comment here: on most road jobs it is possible to work on a number of different sections at the same time, as work sites are needed. On Chapman's Peak, as on the Hex River Poort Sandhills cutting, the construction teams had only two work faces, one at either end. This greatly restricted their efforts.

Finally, on Saturday 6 May 1922 this magnificent scenic drive, planned and built by the Cape Provincial Roads Department, and 'hewn out of the stone face of sheer mountains', was officially opened by Prince Arthur of Connaught, the Governor-General of the Union of South Africa. Tributes were paid to 'Mr Trollip, Chief Inspector of Roads, the late Mr Bromley who had first suggested the feasibility of the road, and to Mr R. Glenday who had surveyed and fixed the line finally adopted and who had been in immediate control of the work of construction'.[12] Prince Arthur rightly described the 11-kilometre scenic drive as 'an engineering feat of the highest order'.[13]

It had taken seven years to complete, and had cost £20,000.

I was amused to come across a handbill, issued by the City Tramways, advertising omnibus passenger trips leaving Cape Town at 11 a.m. round Chapman's Peak Drive, from 8 April 1922 (a month before the official opening) on Mondays, Wednesdays, Fridays and Sundays, for 12s 6d including lunch at the Hout Bay Hotel. The handbill was illustrated with a sketch of an open omnibus – such as one sees in old issues of *Punch* magazine.

An engineering report to the Divisional Council, when they were asked to take over responsibility for Chapman's Peak Road from Province on the completion of construction, mentions *inter alia* that rock falls must be expected, and suitable provision for clearing such falls would have to be provided.[14]

On a road of this nature, places where the drainage measures can be improved show up now and then, especially after heavy rains. In 1959 plans for limited widening and curve improvements were approved, and the work was carried out over the next three years or so. There have also been occasions when subsidences have taken place or washaways occurred: on 14 May 1980 the road had to be closed when quite a large section was washed away and had to be replaced by a bridge-type structure at a cost of R300,000 – it was not possible to re-open the road until 12 December.[15] But, generally speaking, in the intervening eighty years the surfacing of the road has been the only real change. It says something for Robert Glenday's workmanship.

With the emphasis on social and visible services following the 1994 election, funds for road maintenance were cut to below the 'essential' level throughout the Republic. If roads are left without periodic maintenance their rate of deterioration increases exponentially, and the restoration cost increases accordingly. While with lack of maintenance bitumen roads deteriorate early on to an unridable state, it is possible to keep a gravel surface more or less ridable with lesser expenditure. In 1999 the Cape Metropolitan Council realised that with cuts in road-maintenance allocations they would have to face up to the probability of having to close at least ten roads or else return them to gravel surface. One of these was Chapman's Peak Drive. This report was followed by a disastrous fire which destroyed vast amounts of vegetation in the southern Peninsula. As a result there was a loss of the stabilising contribution which the plant roots and vegetation had given to the slopes above Chapman's Peak Drive. The combination of these factors and the tragic death due to a falling rock caused the Drive to be closed in January 2000.

At the time of writing (mid-2002) the Drive is still closed. It appears that the problem is to be handed over to a private concern which, in return for upgrading, will be authorised to charge a toll for the use of Chapman's Peak Drive.

One trusts that upgrading does not destroy too much of the character of the original charming road. Hopefully this wonderful road will soon be opened once again to traffic, not only for the benefit of the local inhabitants, who need it, but also so that we may once again marvel at the views – and at the imagination which conceived such a project and the determination which carried it through to a successful completion.

The first tourists, but not the last
(Hout Bay Museum)

CHAPTER TWENTY-FIVE

BOESMANSKLOOF PASS & BUFFELSPOORT

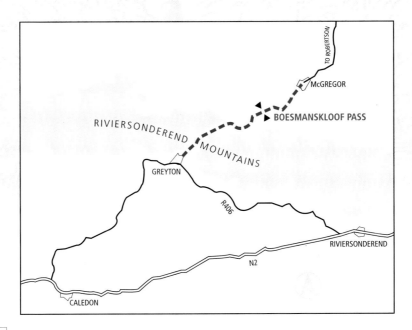

There must be few road engineers who have not got their store of memories of proposed projects which were abandoned for some or other reason, or which just never really got off the ground, despite initial enthusiasm.

As this sort of thing is part of the *padmaker*'s lot, the present chapter deals with 'passes that never were'.

BOESMANSKLOOF PASS

This pass was intended to connect Greyton, on the south side of the Riviersonderend Mountains, with McGregor on the northern side. Greyton is a pleasant little village, lying about 30 kilometres northeast of Caledon. Should the Greytonians for some reason wish to get to the other side of the mountain, their two options are via Floorshoogte, about 50 kilometres to the west as the crow flies, south of Villiersdorp, or via Stormsvleipoort and Robertson, about the same distance to the east.

A footpath exists through Boesmanskloof, showing the presence of traffic between Greyton and McGregor, and so it was decided in 1927 to investigate the possibility of putting a road through the kloof. P.A. de Villiers, later to become well known for his many daring locations of National Roads but at that time a junior engineer with the Cape Provincial Roads Department, was sent to locate and survey a route through the kloof.[1]

Nothing happened for some years – but let J.M. Hoffman, former Provincial Roads Engineer, take up the tale in his inimitable manner.[2]

'About eight or ten years later the Department of Social Welfare thought it would be a good thing to give some of the inhabitants of District Six a chance to enjoy a clean outdoor life and enquired about a suitable site for this exercise. This was where the Provincial Administration hauled out P.A. de Villiers's work of art from the mothballs, Robertson Divisional Council received a grant, a camp was erected to house two hundred labourers with their wheelbarrows, picks and shovels, the labour was recruited and, last of all, I was brought into the picture. The year: 1936 or 1937.

'One morning the District Roads Engineer handed me a roll of plans, explained what was happening, then added that as the labour force was arriving at McGregor that evening under the control of four gangers I had better go and stake some road line. Believe it or not, the original survey stations, although consisting of wooden pegs, could still be located, and by mid-morning of the first day the 200 chaps were busy clearing the formation on the staked line for the first straight.

'Ted Hittersay came to take charge, and I only appeared on the scene again for a couple of days while he was away and then never saw the place again. But in those few days I managed to confiscate a bag of dagga, to have an argument with the District Roads Engineer about my road design, and to keep the police from going into the labourers' sleeping quarters carrying arms. I was young and over-confident!

'Afterwards I heard that the whole job had come to a sorry halt. As a rehabilitation exercise it was not enough of a success – the money dried up and there was excitement about some "phantom" labourers. I suppose someone even questioned whether the road was necessary, and so if you examine the map you will see that there is a road which goes up the mountain from the McGregor side but does not come down the mountain on the Greyton side.'

Greyton and McGregor are both peaceful retreats, great places to unwind. One cannot but wonder what changes to their characters would have happened by now had that road gone through sixty years ago. Today there is a very pleasant hiking trail between the two villages, and I am sure that many a footsore northbound hiker has blessed the few miles of road which Ted Hittersay built all those years ago.

BUFFELSPOORT

Ladismith is in a very similar position relative to Laingsburg as Greyton is to McGregor. The traveller either has to make a wide swing about 50 kilometres to the west to round the toe of the Klein Swartberg before turning back eastwards for 20 kilometres to the

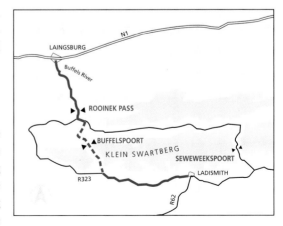

Rooinek Pass on his way to Laingsburg, or else has first to go east for 20 kilometres to Seweweekspoort and then swing back westwards 50 kilometres to Rooinek Pass.

The Cape Provincial Administration felt that as the Swartberg was such a barrier between the South-western Districts and the interior, all possibilities of passes through or over the range should be investigated. The Buffels River runs mainly south and east from Rooinek Pass, and an eight-kilometre poort offers the chance of a more direct route between the two towns. Apparently (I have not walked it) it is a rather tough one, but nonetheless a possibility. So this route was investigated.[3]

First on to the site was a Provincial Roads team, which ran a base-line Distomat survey along the route, through the poort, to which further surveys were coupled. In those days Province appointed land surveyors to do road surveys for us, and this was one of the jobs given out. Besides the normal hazards of a road survey – and in this case the poort was apparently very heavily bushed and forested – the survey party 'had to flee in haste up the side of the gorge one evening when a flood came down. The first land surveyor fell ill in the poort, and had to be helped out bodily to hospital – he later died. A second land surveyor then appeared, and carried out the survey.'[4]

Next on the job were consulting engineers Kantey & Templer. They investigated the Buffelspoort route in considerable detail. Kantey made the best of a very demanding

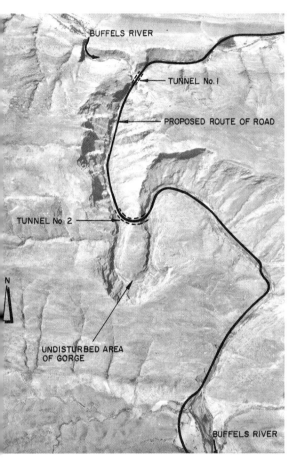

An aerial photograph showing the proposed route (Kantey & Templer)

job. David Bateman has described the poort as 'a similar cut through the mountains to Seweweekspoort'.[5] The need was for a high-standard surfaced road, and when the catchment area of the Buffels River was considered there was no way that the road could run along the narrow bed of the poort. Kantey's designers accordingly had quite a long section where it was necessary to provide precast concrete brackets, as long as the width of the road, bolted to the rock face to keep the road (hopefully) above flood level.[6] This, and other difficulties, obviously made for a very expensive construction.

The route was walked by a number of engineers, including J.M. Hoffman and David Bateman, both very senior in the Provincial ranks. Finally, however, the high cost ensured that enthusiasm for the construction of the route waned, and the project was shelved.

In earlier days one of the attractions of a connection from Ladismith to Laingsburg was the fact that the rail passes through Laingsburg. But in 1925 a branch line was built from Touws River for 153 kilometres to Ladismith, providing direct rail connection. The

These two photographs show the ruggedness and narrowness of the poort (Kantey & Templer)

Laingsburg floods of 1981 damaged the rail line considerably, and thoughts of repairing and reinstating it were abandoned when a review of the transport situation showed that road transport was quite satisfactorily handling the Ladismith demand without the rail backing.

This change in the focus of transport demands, linked to the realisation after the 1981 floods that no road through Buffelspoort could stand up to bad floods, makes it pretty certain that Buffelspoort, as well as being classified as 'one of the passes which never were', can probably also be classified as 'a pass which never will be'.

Buffelspoort was an absolutely pristine wilderness area: it was said that leopards still lived there. Although the poort was apparently completely stripped of vegetation and utterly devastated by the 1981 flood, it has undoubtedly reinstated itself in the interim – nature has amazing healing abilities. I am told that there is some really spectacular scenery in the poort. So if you admire Seweweekspoort, and would like to visit something similar which is off the general beaten path, you may want to put Buffelsrivierpoort on your list of places to be walked some day.

A ROAD WHICH NEVER WAS

In 1951 Prince Albert, at the foot of the Swartberg Pass, was agog because a 'tarred' road had been promised, linking the town to the Laingsburg–Beaufort West National Road. But Prince Albert had two links to the National Road already, the western one running to Prince Albert Road and the eastern to Kruidfontein. Both were the same length, and each had its proponents and its detractors. As I was at that time acting as the local District Roads Engineer, the Provincial Roads Engineer instructed me to locate a route midway between the two.

This involved a considerable amount of veld-riding, the usual amount of walking, and a fair bit of climbing with flags and Abney level while searching for lines over the east–west ridges which crossed our route. It was winter, which is a good time to indulge in this sort of activity.

But the thing which makes this investigation more memorable than other location exercises was riding up the Gamka River from one road to the other, looking for bridge sites. The Divisional Council produced a suitably tame horse, and I was duly hoisted into the saddle with much chaffing and a water bottle, and set off upstream. I had only once before ridden a horse, but we got on very well. I suppose the crow-fly distance would be about 15 kilometres, but what with detours to gates, hitting rock outcrops with my little geological hammer, taking Abney observations and flagging suitable crossings, it was approaching evening when I reached the trees at the other end, where the bakkie arrived shortly afterwards. (I must admit that I am one of those who are quite happy to be alone in the veld, just 'listening to the silence', so possibly the trip could have been done in a shorter time – but then it would not have been so enjoyable.)

In hindsight I sometimes wonder whether the Provincial Roads Engineer was using me to get a decision from the Divisional Council on a preferred route. ('If those of you who want the western route cannot come to an agreement with those who want the eastern route, I'll build it in the middle – where no one wants it!') Shortly after my sorties produced an engineeringly acceptable 'middle' route, the Councillors agreed to go for the western route, and you can ride it today if you pass that way.

CHAPTER TWENTY-SIX

THE STEENBRAS MOUNTAIN ROAD

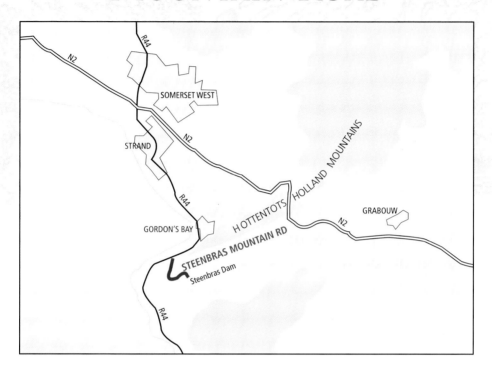

Thus far, when considering the crossings of Van Riebeeck's 'Mountains of Africa', we have not looked further south than Sir Lowry's Pass, behind Somerset West. But there is one more road, behind Gordon's Bay, a purpose-built road providing access to the Steenbras Dam and filtration plant.

The reasons why this pass so frequently gets ignored are threefold. Firstly, it was only built in the 1940s, and obviously made no contribution to breaching the mountain range to open up the country beyond, and so is not mentioned in the histories of the area. Secondly, it is not on any through-route, and so is only used by those with a destination at or near the dam. And thirdly, one has to obtain a permit to enter the dam area at the top.

As you drive east from Cape Town, lift up your eyes and scan the slopes of the Hottentots Holland Mountains spread barrier-like in front of you. You will see the scar of Sir Lowry's Pass striking off to the left, and then swinging round to make its long, steady ascent to the right until it crosses the crest. Now look to the right of that and you will see – from some points looking almost as if it is rising from the waters of False Bay – the mirror-image of the Steenbras Mountain Road, striking off to the right and then swinging left for its long ascent to the summit. And your passengers may comment, when it is pointed out to them: 'Why, I never noticed that pass before!'

The road is surfaced, and obviously engineered. It takes off from the fine Gordon's Bay–Rooi Els coastal road a short distance south of Gordon's Bay and climbs away at a steady 1 in 12 grade, giving fine views over False Bay. It takes advantage of a kloof to switch direction, and then gives fine views over the Helderberg (or Hottentots Holland) valley. The designer has given a short section of flattish grade, to allow you to recover from your 180° change of direction, before resuming a 1 in 12 climb. The road passes the filtration plant (where you buy your permit at the gate, and where there is parking if you want to use

Views of and from the road (Graham Ross)

your binoculars and camera) to climb up another hundred metres to the high point on the saddle. The dam is 50 metres lower down on the far side.

Ronnie Fisk was the engineer who conceived the idea of this link road, proved its feasibility on the ground, and got its construction approved. Let him tell the story in his own words.[1]

'In 1937 an advertisement appeared inviting applications for the post of design engineer for a new filtration plant at Steenbras Dam. At the time I was happily settled in municipal service on the Reef and the salary offered was lower than I was then receiving. However, I possessed experience in the field of water treatment, and getting in at grass-roots level on a project of such magnitude was exciting. A carrot was that a free house on the site was available for myself and my family. I was duly appointed by the City Engineer, Mr W.S. Lunn.

'On assuming duty I learnt that the design team would consist of myself and Mr R.L. Atkins, BSc, AMIChemE, with Mr C. Costley White, the retired Water Engineer and the G.O.M. of Cape Town's water undertaking, as consultant. We were to be directly responsible to the City Engineer but for administrative purposes were attached to the Waterworks Branch.

Looking down towards Gordon's Bay (Graham Ross)

'I was not familiar with the geography of the Cape, and on visiting the dam area I learnt that access was via Sir Lowry's Pass and then along the western shore of the dam on what was in fact a wagon track built during the construction of the original wall. The "free house" was a stone cottage adjacent to the dam wall (it is still there) and access to the plant was over a saddle in the mountain on a track not negotiable by a modern passenger car. [The filtration plant had obviously to be at a lower level than the dam, and was in fact 100 metres lower than the shoulder, as already noted. Access was thus very round-about, to say the least.]

'The first step was to do a detailed contour survey of the plant site while my wife and I camped in the "free house". It was then that I suggested the possibility of a road to Gordon's Bay be explored. I was told that there had been several investigations and that a road was just not economically feasible. I decided that there would either be a road and me, or no road and no me after the design was completed.

'I again approached Waterworks about a road and received a firm "No," so I decided to explore myself.

'Telescopic observations from the plant site and the coastal road were discouraging. I then bought myself a new pair of army boots and two broomsticks to which I nailed cross-pieces at eye level. I borrowed an Abney [level] and surveyor's chain, five labourers and tools from the forest ranger, and camped in the "free house".

'The first day we cut pegs and set out supplies along the route that, from the start, I thought the road might follow. The second day, after setting the Abney for 1 in 12, we hit the side of the buttress that had been considered the greatest obstacle, and traversed over it. That night, after plotting the traverse, I knew that I had struck gold. By the next night my boots had worn through and I had to return to the City.

The road winding uphill (Graham Ross)

'The following week there were no real obstacles except for a rock face that was both high and long and would need a high retaining wall for a road to bypass it, and a deep ravine that was actually lucky because it would accommodate a traffic circle for a 180° turn. Back at office the traverse was plotted, as were cross-sections at all the pegs, where angle of slope up and down had been taken.

'Came the day for the final inspection, and on the wall of my office was a large-scale coloured-in drawing of the layout, at the top right corner of which were dotted lines and "Possible road to Gordon's Bay". On seeing that, Lunn's eyes lit up and he said, "Can you get a road to Gordon's Bay?" I said, "Yes." He then asked, "What grade and what width?" I replied, "1 in 12, and 22-foot formation." He asked, "Why?" and I said, "Because that is Sir Lowry's Pass at its narrowest." "What would the cost be?" I gave an estimate. He replied, "Build it, but 26-foot formation!" and walked out, followed by some disgruntled-looking Waterworks officials.

'By shortening the distance from the plant to Gordon's Bay, the road has today paid for itself several times over.

'And it also provides views that are beyond price!'

Well, it nearly was 'a pass that never was', and but for Ronnie Fisk's doggedness would definitely never have come into being. If you look at a map of the area you will immediately see what a saving in distance the road provides. And don't forget his filtration plant at 300 metres elevation was 100 metres below the 400-metre-high saddle (1.2 kilometres along a 1 in 12 road)!

I have travelled that road, and it is everything Ronnie said it is. The construction is interesting, and the views are, as would be expected, stupendous. It is well worth visiting.

The 180° turn halfway up (Graham Ross)

CHAPTER TWENTY-SEVEN
DU TOIT'S KLOOF PASS

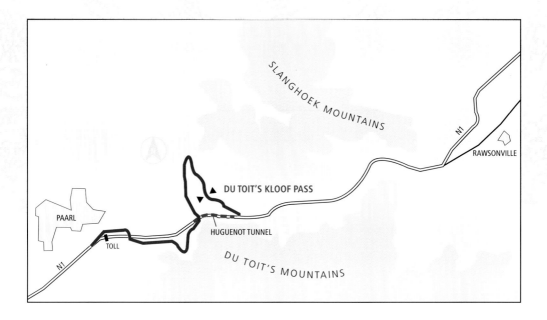

Du Toit's Kloof Pass, 50 kilometres east of Cape Town on the N1 to Worcester, Johannesburg and Zimbabwe, is strangely a Johnny-come-lately among the Cape passes. This can be largely ascribed to the fact that for many years the Great North Road ran via Ceres and Sutherland, so that efforts were understandably concentrated on Bain's Kloof and Michell's Pass.

It was only in 1936, with the promulgation of the first National Road five-year plan that people came to accept that the emphasis should switch from the Ceres route to one via Worcester. And it was not until 1940 that agreement was reached that the route should go through Du Toit's Kloof rather than via Wellington and Bain's Kloof.[1] Undoubtedly the difficulty of improving the road through Bain's Kloof to the standards needed for modern traffic was a weighty contributory factor.

Du Toit's Kloof had been used to get from one side of the Hawequas to another, skirting the toe of the Klein Drakenstein Mountains, from early times. Undoubtedly wild animals made the first transit. Later the nomadic Khoi stock farmers used what was known as the Hawequa cattle track when the grass was greener on the other side, as was recorded by François du Toit, who commenced farming on the Paarl side in 1692.[2]

Simon van der Stel wanted to ease the lot of the transport riders who fetched timber from the Riviersonderend valley. In his advice to his son, Willem, when the latter succeeded him in the post of Governor in 1699, he mooted a road in Du Toit's Kloof, over what he called 'Het Olifants Pad'. He estimated that the construction time would be three months. Unfortunately Willem did not follow up on this suggestion.

Joshua Joubert built a road through the kloof in 1738. He got very cross when others and their cattle used his private road, and in 1785 petitioned the government for permission to levy a toll, also asking for financial assistance to help him finish the job. The government was not sympathetic to either request.[3]

That indefatigable traveller Carl Thunberg mentions that a track practicable for horsemen existed over the mountains at 'Jan de Toi's Kloof' in 1772.[4] William Burchell, in writing up his travels and including helpful hints for future travellers, says: 'Since my departure from the Cape a convenient wagon road has been made over the mountains near the Paarl, at a place called Du Toits Kloof, which till then was merely a footpath.'[5] He was well intentioned but misinformed: at that time the wagon road had not got past the talking-about-it stage.

But every now and then there seemed to be someone trying to improve the tracks through the kloof. Lieutenant Detlef Siegfried Schonfeldt, who had bought a farm in the kloof, drummed up some financial contributions from private citizens, and tools and explosives from the government. He managed to build a road of sorts up to and over the neck, and down the kloof on the eastern side. After two years he was finally beaten and ran out of money when trying to make a safe road over the Kleygat (through which the first National

Under construction, circa 1948 (Transnet Heritage Library)

Road tunnelled), which was the most dangerous part of the track. In 1853 Dr Gird 'recollected the terrible dangers of Du Toit's Kloof, where the path lay over the steep face of the mountain at the Kley Gat, where one false step would plunge the traveller into eternity'.[6] In 1826 Schonfeldt petitioned the government for assistance: he wanted £1,600 with which he reckoned he could pay 60 labourers who would finish the road in three or four months. The project was reviewed by Lieutenant Charles Alexander, Royal Engineers, who reported on 6 May 1826 that 'although the pass was not practicable for waggons, Mr Schonfeldt had with considerable labour and expense made a safe and tolerably convenient thoroughfare for cattle. The improvement is particularly marked at the Kleigat. Although some of the most difficult spots have been rendered passable by waggons, it would require an expense of two or three thousand pounds sterling to make the pass equal to the general mountain roads of the Colony, inferior to Fransche Hoek but superior to Hex River Kloof.'[7]

This same Alexander had reviewed the costs of the Franschhoek Pass the previous year, when the Governor, Lord Charles Somerset, had been reprimanded by the Colonial Office for approving works with costs exceeding his delegated powers-to-approve. It was not the best time to request financial assistance: poor Schonfeldt's application was turned down. Although the Council approved (10 May 1826) and the Local Ordinance of 31 August 1826 promulgated the right 'for the said Detlef Siegfried Schonfeldt, his Heirs and Assigns' to levy a toll to cover the cost of repairs, for a period of five years, as a measure of recompense for his previous expenditure, the unfortunate man, having spent his own money on a project in which he had faith, was ruined.[8]

Without adequate maintenance the road deteriorated to such an extent that only six years later, in 1832, the *Cape of Good Hope Almanac and Directory* warned: 'Du Toits Kloof, which

Western tunnel portal under construction (VKE Engineers)

had been made practicable for wagons for a certain distance, by the private exertion of Lt. Schonfeldt, is now only passable for cattle, foot passengers and horsemen, which latter are, however, obliged to use much caution, and to alight at several places; some fatal accidents have of late occurred on this pass.'

Major Charles Michell, in his comprehensive paper read before the Royal Geographical Society in 1836, had this to say of Du Toit's Kloof: 'the great length – about ten miles of very difficult rocky road – and the expense ... induce me to regard it more as a work to be achieved by our posterity than within the probably available means of our own time.'[9] And so it was.

As a result of petitions received – perhaps in opposition to the proposal which was made

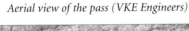

Aerial view of the pass (VKE Engineers)

at that time to build Bain's Kloof – Charles Michell and Andrew Geddes Bain in 1846 carried out a considerable investigation and survey of the Du Toit's Kloof route. They estimated that the cost would be £340,000, which would have to have been paid by the inhabitants of Paarl and Worcester. This was manifestly impossible, so the proposal was dropped. In 1858 Captain George Pilkington, who succeeded Michell as Colonial Civil Engineer in 1848, chose another line, for which he gave an estimate of £45,000, but even this cost was too high for the ratepayers. Nothing much happened for quite some time after this.

The next noteworthy action was in 1929, when the Town Engineers of Paarl and Worcester teamed up to survey a route through the kloof. This reduced the distance between their towns by 15 kilometres. Their estimate of the cost was £30,000 to £40,000. No action resulted, but the pressure was mounting.[10]

P.A. de Villiers was involved with Du Toit's Kloof at intervals throughout his career. It was he who promoted the idea of a road through the kloof while working for the Cape Provincial Roads Department, and in 1932 actually persuaded the Chief Inspector of Roads, Mr F. Beck, to accompany him on a walk through the kloof. Although the first National Road five-year scheme in 1936 adopted the route through Worcester and the Hex River Pass, the road still ran via Bain's Kloof. De Villiers was appointed National Roads Inspecting Engineer in the Western Cape in 1937, and was apparently a moving spirit in getting Du Toit's Kloof accepted in preference to Bain's Kloof. Even at that early stage he was suggesting a tunnel to reduce distance and the length of rise and fall along the road.[11] P.A. liked tunnels! When he retired, consulting engineers Van Niekerk, Kleyn & Edwards (VKE), the local engineers who, together with Electrowatt, carried out the design and supervision of the Huguenot Tunnel, appointed him as a consultant, so he was able to retain his contact until the end.

During 1940 P.A. ran a survey through Du Toit's Kloof and prepared a report with an estimate of £360,000. This was the location on which the alignment and design were based when the road was constructed. But South Africa was at war then and most of the *padmakers*, engineers and support staff alike, had volunteered for service and were active at points north. So nothing could be done – until the Roads Department was approached in 1941 to provide occupation for some of the thousands of Italian prisoners of war who were interned for the duration.

Four or five hundred POWs were sent to Du Toit's Kloof. No Provincial engineer was available as Resident Engineer, so an experienced foreman, D. Coetzee, was put in charge, with occasional visits by P.A. de Villiers.[12] When the war ended in 1945 and the Italians were repatriated, local labour took over and in fact constructed most of the pass. The job was 46 kilometres long from the Berg River bridge to the Goudini junction, with a 7.3-metre carriageway, and cost £750,000.[13] A tunnel through the Kleygat nose finally sorted out this troublesome feature. It is recorded that the 222-metre 'road tunnel built under the direction of Resident Engineer M.C. Vos was excavated in approximately three months by a crew that had never tunnelled before, but Vos did buy and study a book on tunnelling before starting the operation.'[14]

J.M. Hoffman, who took over from P.A. de Villiers as National Roads Inspecting Engineer in the Western Cape and was later the Cape Provincial Roads Engineer, has a story about the tunnel. 'I remember walking into the tunnel while it was under construction in the company of the Construction Engineer, Eric Fergus, and Works Foreman Dick Coetzee. Fergus was stressing the importance of checking the ceiling for loose rocks and Coetzee was assuring him that he did that personally after every blast. Just then a sizeable dislodged rock fell behind Fergus, but what with the roar of the jack-hammers at the face he did not notice a thing and I did not think it necessary or desirable to mention the matter.'[15]

The majority of the work on the pass was done by a Provincial construction unit based near Worcester, but the last ten kilometres on the Paarl side were constructed by people at a sub-camp there. On site there was also a Landscape Officer, trained horticulturist Bill Sheat, to oversee the restoration of the natural flora.

The pass was opened by the Prime Minister, Dr D.F. Malan, on Saturday 26 May 1949. We were told to keep off the road and out of sight of the dignitaries attending the opening ceremony. We congregated at the sub-camp, which was alongside the road at the Berg River bridge. There I learnt something to put in my book on staff relations: if for reasons of space, or for any other reason, it is necessary to exclude from the opening ceremony those who have worked on a project, it is a good idea to arrange a braai or some similar happening instead of just herding them away to growl. However, all went well once I had persuaded one of the staff, who was leaning over the sub-camp fence saluting each passing black limousine with a raised right hand and a muttered 'Up your pass', to return to the sub-camp buildings with the rest of us.

Although the pass was finished in 1949, it was not until 1951 that the new National Road link between Paarl and Cape Town was completed.[16]

Traffic built up, and the 50 k.p.h. design speed on the upper sections of the pass and the resulting no-passing zones began causing traffic back-ups. Such palliatives as were possible, mainly the construction of additional up-and-down crawl lanes wherever space was available, ameliorated the problem but in the long term were inadequate. Finally, the road authorities bit on the bullet, and appointed the VKE consortium to carry out the essential geological and other investigations, to design and to supervise the construction of the four-kilometre-long Huguenot Tunnel and the dual-carriageway approaches on the western or Paarl side. This project was opened in 1988 and, as is appropriate for such a monumental work, has been well documented in the technical and lay press. So I am not going to write up any more about the tunnel, except to say that I went through it, suitably clad in sea boots, oilskins and sou'wester, when it was under construction, and my comment is that I would rather work in the open air.

The tunnel is doing its job very well. When last I heard, the accident rate was 179 per 100 million vehicle kilometres, compared with the national average of 460.[17]

The approaches on the Worcester side needed attention in due course, and on 19 June 1997 the Minister of Transport, Mac Maharaj, opened the 13.6-kilometre dual-carriageway road from the tunnel to Florence. Construction took four years, and the cost was R125 million.[18]

Du Toit's Kloof Pass is a very fine piece of engineering, and we may well be proud of it. But the view from inside the tunnel is limited, and I still take my caravan over the top on the old road – and I have yet to be able to do the trip without pulling off in a couple of places to look at the scenery.

CHAPTER TWENTY-EIGHT

GREAT BRAK PASS

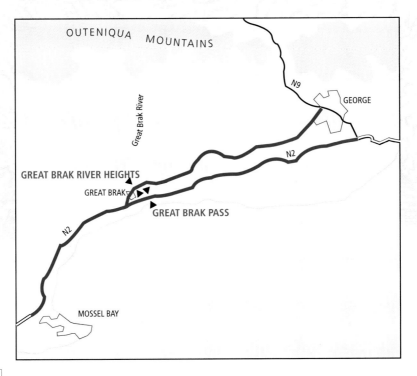

In days past, the escarpment on the eastern or left side of the Great Brak River was the main physical impediment encountered on the road connecting Mossel Bay and George, in the Southern Cape. It was not a really serious impediment, compared to some of the passes we have considered, but was high enough and steep enough to require considerable effort to drive a wagon up the slope, and to cause some excitement when descending with a wagon.

There was of course traffic between Mossel Bay and George before an engineered pass had been constructed. August Frederik Beutler came this way in 1752, followed by Joachim van Plettenberg in 1778 and Johan Victorin in 1854, among many others.

GREAT BRAK RIVER HEIGHTS

But by that time an effort was being made to improve the route between the two towns. Thus, in 1844 a wooden bridge was built over the Great Brak River by convict labour under the supervision of Resident Magistrate Moodie of George.[1] And Henry Fancourt White, fresh from building the Montagu Pass (opened in January 1848), tackled the construction of the Mossel Bay–George road. In November of that year he was able to report to the Central Road

Board: 'The New Pass at the Great Brak River Heights has been lately opened for the use of the Public ... The old road was so steep that it was a work of toil for an unloaded animal to ascend it. On the new line ... the heaviest loaded wagons have gone from the top to the bottom without locking a wheel; and a single horse with a gig may trot either up or down the whole length ...'[2]

Victorin used the bridge and the new pass in 1854. He recorded: 'the Great Brack River ... was crossed by a wooden bridge. Immediately after that "The Hills" begin, about a 2-mile long climb to the plateau where George is situated ... The road to the plateau is extremely beautiful and also quite good, although there were few places where one could hope to escape with one's life if the wagon overturned ...'[3]

White's road from George was near the foothills. His Great Brak River Heights is thus not where the National Road runs today but a few kilometres upstream, above the village of Great Brak. The river is crossed within the village itself. Although undoubtedly improved over the years, with a major revamp in 1933,[4] the pass still rises 300 metres in three kilometres, and runs through most attractive country. This gravel route makes a pleasant change from the hurrying, scurrying National Road if you are on your way to George.

Moodie's 1844 bridge served a real need: in December 1849 some 193 wagons and 30 carts crossed here. Unfortunately it was not very well designed or constructed, and it had to be replaced in 1850. The Colonial Civil Engineer, Captain George Pilkington, provided a design which incorporated a causeway with 13 stone piers, between which twelve 20-foot wooden decks floated. This design apparently came in for a lot of flak ('a pretty notion, very

Bain's map showing Great Brak River Heights on the way to George (Cape Archives)

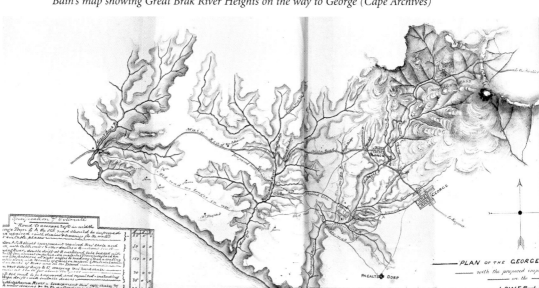

fine indeed in theory; only the worst of it was that the theory could not be reduced to practice'). Nonetheless, and despite these forecasts, it did work in practice – until 1965, in fact, when the floating decks were replaced by a solid bridge structure. And that deck was founded on Pilkington's 13 piers.[5]

Communications were opening up throughout the country as improved roads made the use of lighter, horse-drawn and, hence, faster vehicles possible. By 1870 the post-cart (which carried a few passengers as well as the mail) 'did the journey from Cape Town to George in 48 hours, which meant travelling for two days and two nights with nothing that could be called rest, except for sitting down for a quarter of an hour at a farm or wayside inn, morning, noon and night for a hasty meal'.[6]

GREAT BRAK PASS

But the standards of the old, pleasantly meandering road from Mossel Bay to George were not acceptable for the National Road, which connected the two towns in the 1950s. Another east–west road was at any rate needed to serve the coastal shelf there. So a totally new line was developed, and a new pass was constructed up the escarpment to the east of the Great Brak River. This construction was carried out by the Provincial Roads construction unit based at George, which also built the Outeniqua Pass. It incorporated a new bridge over the river, next to the railway bridge, and was a great improvement from a traffic-carrying point of view.

In accord with the traffic volumes of the time, it was built with a single, two-lane carriageway. Then came the bold policy decision that all National Roads should be freeways. What is more, traffic was building up, especially along the 'Garden Route' and in peak seasons.

In this vicinity the first freeway construction was the bypass of Mossel Bay, providing also a shortening of the travel distance for through-traffic. Next, the freeway was built over the plateau to the George airport, later extended as a bypass of George. This left the Great Brak Pass section and its approaches, where a four-lane dual-carriageway road built to freeway standards had to be fitted in. It was quite a challenge.

The really niggly bit was the river crossing. The railway bridge was (of course) in the optimum spot. There was also the existing road bridge, which had to be left for local access. If one crossed upstream of those two, one would be headed for the slope of the escarpment, which swung around to the east: a very big side-cut. The logical place was therefore on the sea side. But then someone found out that a colony of very special and unique prawns was located just where the road had to go.

So it was back to the line upstream of the old road bridge, with all its problems. One of these was that one side-cut would literally extend right to the top of the plateau, creating a massive bare slope which would present a horrible scar on the landscape unless properly treated. There were also other cuts, large when considered individually, which had to be dealt with. Now, *padmakers* are well known for their environmental sensitivity; in fact, they were environmentally sensitive long before the word 'environment' became so popular. But for this job specialist specialists were called in to back up our team – and the end result justifies the care and consideration given it.

This massive side-cut became quite well known. Some time after the road was opened, a number of engineers from Port Elizabeth, wishing to see and learn for themselves, hired a bus and came down one Saturday. Standing on the roadside they had to crane their necks, and couldn't really see much. So they got into their bus again and drove a kilometre or so off to the side, to the beach. Imagine our pride when the only way they could tell where the cut had been made was by association with the river bridge. The care and planning put into the re-vegetation had paid off.

The pass climbs from five metres above sea level to over 200 metres in about eight kilometres to the Glentana Interchange on the plateau. Andrew Tanner was Ninham Shand's Resident Engineer on the contract, and has prepared a monograph, covering some of the aspects of the job which stuck in his mind.[7]

'Construction of this section of the N2 between the Great Brak River and Glentana commenced in October 1977 along a route designed to blend in with the topography. One of my early memories was sharing the platform, with a good deal of trepidation, with Martin Blumerus, Clifford Harris's Contracts Manager, at a presentation to the Great Brak Residents' Association to explain what we were about to do to the rolling hills above their village. As is often the case, the rumours that preceded us about the destruction we were about to wreak were far worse than the reality, and our being able to reassure the residents with definite information paved the way for an amicable relationship throughout the contract. Further positive public relations initiatives included site visits for matric students from the local high school, the Southern Cape Informal Branch of SAICE, and interested residents.

The first Great Brak Pass, circa 1963 (Graham Ross)

'Relations with the land owners were generally cordial, but one farm was dissected by the road and the road passed within fifty metres of the farmhouse. This was occupied by two formidable gentlemen, the Young brothers. Neither was particularly young, but both were very clear about what would or would not be acceptable. What was not acceptable was the contractor moving onto their property before they had received their money for compensation, and the vivid account of the contractor's operators, who came face to face with a double-barrelled shotgun one morning, prompted very speedy action through the hierarchy …

'On a construction project of this magnitude, R15 million in 1981, numerous incidents come to mind. The majority relate to the very large cuts that formed part of the contract, three with slope lengths of the order of ninety metres. The third one after crossing the river bridge was known as the 'granite cut' and the slope design provided plenty of interest. The stability was controlled by joints that could only be properly assessed during major excavation and a pilot cut to the full depth was specified and executed. Careful measurements and analysis led to the final slopes being determined and "final" rock excavations proceeded …

'As the final cut-slope was excavated, the face was covered with vegetation cylinders to assist topsoil retention and promote the growth of vegetation. On driving through the site one day I noticed that at one end of the cut the rows of vegetation cylinders were not horizontal as required, and I instructed an Assistant RE to inspect the problem. While standing at the bottom of the slope he heard a strange noise, which he described as rather like a loud snake hiss, and found the cut-slope coming closer to him. He had just observed, first hand, a minor slope failure, which had also been the cause of the strange angle of the vegetation cylinders. The slip revealed a new series of planes of weakness, flatter than those which had previously been observed, and a major re-design of the eastern end of the cut-slope was required. The cut-slope was re-examined, redesigned and re-excavated. This was no mean task, necessitating construction of a road up to the top of the cut and starting again.

'Finally the cut-slope was finished, the grass all planted and thriving. To our horror one day it appeared as if a giant lawn-mower had moved across a portion of the cut-slope, mowing the kikuyu down to a short stubble, with some areas rapidly becoming brown. Closer inspection revealed an infestation of army worms, which were then suitably dealt with.

'Throughout the contract considerable attention was paid to protecting the natural vegetation and determined efforts were made to re-establish the fynbos and various experimental approaches were tried. A large vacuum cleaner, nicknamed 'the billy goat', was used to vacuum up leaf debris and, it was hoped, seeds from the areas of natural fynbos, and this was then incorporated into the seed-and-fertiliser mix that was hydroseeded onto the slopes. However, the tankers used for the hydroseeding were not accustomed to such fibre-rich diet and the pumps immediately clogged. Plan B involved hand-spreading the harvested leaf-and-seed mix, and considerable success was achieved.

'After negotiating four major cuts, the route crosses a short plain dissected by three dongas up to eighteen metres deep, before commencing the final climb onto the plateau. Access to the bottom of the dongas for construction of the culverts was quite tricky, while the main haul road through the site crossed the dongas upstream of the culverts. In the deepest donga, immediately upstream of the culvert, was a waterfall, with the haul road at the top. One afternoon when the final concrete in the base slab of the culvert had just been poured, a truck driver decided that he had to respond to a call of nature and parked his truck on the haul road above the waterfall. The truck started to roll back towards the edge of the waterfall while the concreting gang was carrying its equipment into the site hut at the foot of the waterfall. Seconds later the truck landed on the just-completed culvert base-slab, "parking" right outside the door of the hut. The concreting gang peered fearfully out of the hut at the truck now embedded in the soft concrete where they had been working only moments earlier ...

'Although the recollections that come to mind most readily and that are, with hindsight, the more amusing are of problems, the overriding memories of the project are of successes achieved and difficulties overcome. A good team spirit developed on site and the contractor's and consultant's staff were often seen at lunchtime enjoying a run on the beach and a swim, among other joint social activities. After three years of hard work by a dedicated site team from contractor, consultant and client – many of whom are still friends – the road was officially opened on 6 May 1981 by the then Prime Minister, P.W. Botha. It stands as a landmark in engineering design, closing the gap between two previously constructed sections of this 41-kilometre portion of the freeway (one along the coastal dunes to Mossel Bay and the other across the plateau to George), providing magnificent views across the bay with major cuts and embankments that today blend harmoniously, if somewhat geometrically, with the surrounding hills.'

When next you drive Great Brak Pass, give some consideration to all the thought and care which went into the design and construction of this project. Of course, this applies to all civil engineering works, only undoubtedly more so when the work is a road mountain pass. May you enjoy the artistic symmetry of the completed passes you travel, and may they speed you on your way. Just as the Great Brak Pass does.

A massive cut on the 1981 Great Brak Pass, with the old two-lane pass lower down on the left (Andrew Tanner)

CHAPTER TWENTY-NINE

OUTENIQUA PASS

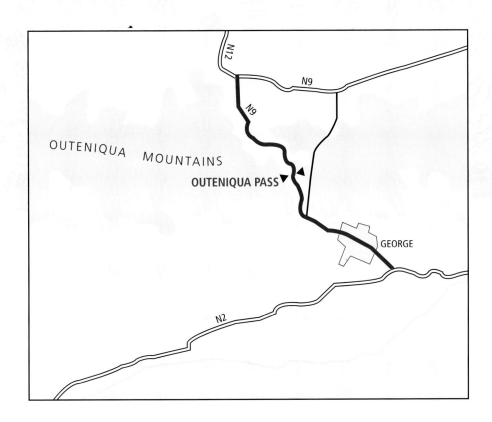

The historic Montagu Pass links George and the coastal plain with the Little Karoo north of the Outeniqua Mountains. Designed by Charles Cornwallis Michell and constructed by Henry Fancourt White, it was opened in 1847. For over a hundred years this monumental pass carried the traffic safely over the mountain range, climbing the eastern slopes of the Klip River valley. But the ever-increasing traffic volume finally demanded the provision of a facility with greater capacity.

It was not feasible to widen and otherwise improve Montagu Pass itself to the required standards. It was also not easy to find a location with acceptable standards. Over a number of years various engineers investigated alternatives, even considering routes in the vicinity of Robinson Pass behind Mossel Bay. In 1937 P.A. de Villiers, then Location Engineer with the

National Road Board, sought and then developed a route down the western slopes of the Malgas River valley, after considerable field investigations which included a reconnaissance flight over the area. His proposed alignment was eventually accepted for construction.[1]

This route, like most pass alignments, was not without its snags and challenges, but with modern road-building expertise and machinery it was possible to construct a pass to acceptable standards. To satisfy the usual contrasting demands of adequate standards and economy of construction, a seven-metre carriageway, with a maximum gradient of six per cent and a minimum curve radius of 90 metres, was decided upon. To satisfy these requirements the topography of the mountain required that there be rock cuts 20 metres deep and fills more than 30 metres high, but environmental damage was kept to an unavoidable minimum.[2]

P.A.'s original line included two short tunnels at the summit. Reuben Stander, then Assistant Resident Engineer on the unit and responsible for pegging out the line for construction, came up with a proposed realignment which eliminated the necessity for constructing the tunnels. After detailed investigations, additional surveys, much design work, and inspections *in loco* by P.A. de Villiers of this and a later alternative by J.M. Hoffman (then National Roads Inspecting Engineer), the road was finally built without tunnels over a neck, 60 metres higher than the original route.[3]

In 1942, when construction of the pass began, South Africa was at war and most of the *padmakers* were up North in the South African Engineering Corps road units. The country also had to accommodate and find occupation for a large number of Italian prisoners of war. Some of these POWs were sent to George, to work on the construction of the pass. Opinions vary as to exactly how many were there: one source says 800 and another 500, while Pieter Baartman, a surveyor working on the pass, cannot recollect such a large number. His guess is that there were about 200 actually working on the pass.

Be this as it may, construction started in October 1942, with design, supervision, surveying and materials control being provided by the Cape Provincial Roads Department. The POWs, who included in their number professional musicians and artists but nary a roadbuilder, provided most of the labour until they were repatriated. Progress was slow, prompting quips about Rome not having been built in a day.

Pieter Baartman has again favoured us with various reminiscences about this construction period.[4]

'When first introduced to the Outeniqua Mountains in January 1944, I was 16 years old and freshly matriculated. The earthworks construction of the new pass, which would replace the historic Montagu Pass, had only penetrated the foothills and work had barely started on the first really large embankment. The mountain massif lay ahead, and during the following five years I would be very much involved with this daunting project until the construction reached the summit and work on the northern end of the pass had been commenced by a private firm of contractors ...

'My immediate bosses were the Unit Surveyor and the Assistant Resident Engineer (ARE), who had graduated the previous year and started work a month before me.

'I had hardly settled in before my routine duty of *handlanger* to these two men commenced. We left camp each morning at 6 a.m. aboard our survey van, at that time a Chevy one-and-a-half tonner (painted the standard red of all National Road Board vehicles), and headed for the mountain ... I discovered that close to Witfontein was an Italian prisoner-of-war camp from which willing prisoners were recruited at a wage of one shilling per day to work as labourers on the pass construction. I came to envy them their daily ration of hot soup, transported from Witfontein in a huge cauldron, while I munched my dry sandwich, washed down with cold black coffee.

'When I accompanied the Surveyor to the construction site, it was to assist him with the replacement of damaged or destroyed line and level reference pegs or with the monthly task of measuring the quantities of earth and blasted rock moved and concrete cast. After this field excercise I was taught to process the field observations and calculate the figures needed for cost accounting. If ever I learned patience with drudgery, it was by performing this job every month for four years.

'My duties as assistant to the ARE were far more satisfying. He was responsible for pegging the line of the road ahead of construction and producing the working plans, in cross and longitudinal section, on which the vertical alignment of the road was designed and the proposed quantities of excavation and embankment calculated. I started off as a labourer, pulling the tape, driving in the pegs, cutting paths through the often dense mountain foliage, carrying water and cement, and building the small concrete monuments that protected the steel reference pegs. For the first year of my service I was not permitted to handle any of the survey instruments other than, after learning by observing, to carry them between set-up stations. I was made to understand that if any harm came to the theodolite that made it unusable, the work on the construction would virtually grind to a halt and the perpetrator of the deed would undoubtedly be dealt with most severely, or worse! That, if anything, made me realise how vital was the job of the surveyor and I became convinced that by carrying the instruments at exactly the correct angle, I was not only protecting my own livelihood, but ensuring the future of my career.

'The topo plans of the route we were pegging out indicated a proposed tunnel. I had overheard a fair amount of discussion about this project and the mastermind behind it. This gentleman was held in considerable awe: he was known as 'the Tunnel King' and had also been responsible for the location of the Du Toit's Kloof Pass route and its tunnel.

'One day the ARE left me establishing reference pegs while he tramped off towards the summit of the mountain. He returned several hours later looking very pleased with himself. Next day, accompanied by two labourers carrying bundles of flagged poles, we set off, myself ahead with the flag-carriers while, bringing up the rear, the ARE, using a hand-held Abney level, directed us onto the line of a permissible road gradient ... We returned the following day with our full survey party and carried out a topo survey along the new route and plotted the plans. After the ARE had drawn in his alternative design, he presented it to the RE for comment and, if approved, for sending off to Pretoria for their consideration. Within days the Tunnel King arrived, looking rather grim, I thought. There was much poring over the plans, sometimes voices were raised, and then all concerned went off to the mountain, plans in hand, for a site inspection. The tunnel was abandoned.

'However, the ARE was not done yet. He came up with another proposal – the final approach to the summit should be re-investigated. Instead of the summit crossing as planned, an alternative route should be examined that continued along the southern face of

Outeniqua Pass on the left and Montagu Pass on the right (Kelvin Saunders)

the mountain to cross at a different summit close to the little settlement of Herold. Head office requested a detailed report on this proposal, and again I accompanied the ARE on a location expedition, this one more extensive than the tunnel alternative. Before any further progress was made, the ARE resigned from the Department and I was left as the only person with physical knowledge of the actual line of the proposed route.

'In due course we were notified that a Very Senior Engineer fron the Pretoria office would arrive to assess the alternative route and I was to be available to point it out. The Surveyor gave me a very serious lecture before the event: I should be aware of the honour I would

Outeniqua Pass and the Outeniqua Mountains (Kelvin Saunders)

enjoy in being in the company of this illustrious gentleman, I was to pay close attention to every word he uttered, and every gem of wisdom he offered should be remembered. I was to dress as well as possible, not forget to brush my teeth or comb my hair, and take along a clean handkerchief. It seemed my whole future with the Department was in the balance. If I performed well and absorbed all the knowledge imparted, I would never look back.

'Three of us, the RE, the Great One and I, set off on our expedition. After about half an hour the RE was forced to drop out and find his way back to the car. His ankles could not take the strain of the climb and in any case he was obviously in no physical condition for this type of exercise. My remaining companion was a totally different kettle of fish. Fit as a fiddle, he had no problem keeping up and all the while he regaled me with marvellous anecdotes, none of which had anything to do with the task at hand. One tale involved some strange trees (found only on the Scilly Isles) that, like pawpaws, came in both sexes and apparently did remarkable things at times of full moon. Many years later, this fine gentleman and great friend attended my retirement function, where I took the opportunity of reminding him of this interlude and the lecture I had received. I regretfully had to tell him that our association on that day had done absolutely nothing to further my career!'

Let J.M. 'Hoffie' Hoffman continue the story.[5]

'The Du Toit's Kloof and Outeniqua Passes were commenced during the war years ... in order to assist with the problem of employing the large contingent of Italian POWs that had been brought to South Africa. About 500 went to Keerweder on the Du Toit's Kloof job and 500 to George for the Outeniqua Pass. Now, Italy is known for its mountain passes, and by association the legend grew that these two passes were built by Italian prisoners of war, implying at the same time that such work was beyond the competence of South Africans. While I was Provincial Roads Engineer I received an enquiry from the Italian Embassy about the contribution of their compatriots in the building of these passes. I paged through our files to see whether I could find any suitable information and came upon a letter from a certain Resident Engineer requesting that the 500 POWs please be taken away and replaced by 250 Africans so that he could get on with the job. This was not considered suitable information!

'Outeniqua Pass was a tough job. The drill bits barely penetrated three inches before they had to be resharpened or replaced, and drill sharpeners were in high demand. Progress was slow, and when it was suggested that from the summit northwards a section should be contracted out, no one objected.

'Clifford Harris was awarded the contract and a few days later phoned and enquired as to what convenient railway siding he could consign a couple of large self-loading scrapers. He had them spare and had to park them somewhere. Well, to cut a long story short, he did practically all the earth-moving with those scrapers. Needless to say, to our envy and dismay, there was no hard rock on his contract, certainly not rock of the quality encountered on the departmental works and expected on the contract section!'

At the end of the war in 1945, about one-tenth of the work had been finished. The understandably unenthusiastic Italians were returned to their homeland, and their place was taken by African labour. It is said that the pace of construction picked up at once.

A most interesting aspect of the construction is that no culverts were provided under the high rock fills: the brave decision in the late 1940s that percolation through the coarse rock would be adequate for drainage purposes has been proved to be correct.

The pass section is 14.5 kilometres long, and the construction cost approximately £500,000. The pass provided added safety and capacity at the expense of increasing the road distance from George to Herold by 13 kilometres: a good bargain in anyone's language. It was opened on 20 September 1951 by Minister of Transport Paul Sauer.

♦♦♦♦♦♦♦♦♦♦

Fifty years have passed since that day. The volume of traffic continues to grow, as it has a habit of doing. By the 1980s the increasing queuing caused by the lack of passing opportunities on 80 per cent of the pass, and signs of distress in the pavement, indicated that drastic action was once again advisable.

Accordingly, in 1988 the Provincial Roads Engineer appointed Kantey & Templer as consulting engineers for the project. K&T investigated all the complex engineering aspects, and Hill Kaplan Scott assisted by making an environmental impact study. This thorough investigation produced a proposal which, again, balanced improved standards, environmental protection and economic restraints. It was decided in 1990 to design and reconstruct a

13.5-kilometre stretch of the pass proper to the recommended standards – lack of funds unfortunately prohibited upgrading the abutting sections to the north and south.

Construction commenced in September 1993, and LTA Earthworks South executed the contract in the spirit of the environmentally sensitive design which had been prepared. Widening existing cuts and fills by a few metres is always fiddly work. On this contract, not only did it have to be done under traffic, but damage to the unique natural fynbos vegetation had to be avoided wherever possible. Approximately 185,000 cubic metres of hard rock cut-slopes, sometimes up to 30 metres high, had to be blasted without scattering rock far and wide, and the resultant rock debris had to be removed from the roadway as speedily as possible to allow traffic to resume. This rock then had to be placed in 210,000 cubic metres of narrow fill spread along the flanks of existing fills.

The provision of passing lanes, improvements to the pavement, and the use of gabion structures and stone masonry walls have resulted in a project which not only has considerably enhanced the capacity and engineering strength of the pass, but also fits naturally into its surroundings.

The traveller will admire the visually appealing construction aspects, and be grateful for the freer traffic flow and safer operation resulting from the improved passing opportunities. He or she will also undoubtedly appreciate the provision of a number of attractively designed formal parking and view sites, which allow one to stop safely to admire the stunning natural beauty of the surroundings.

The work was spread over four years and cost R62 million, of which R3 million was directly attributable to environmental controls and vegetation. The revamped pass was re-opened by Western Cape Premier Hernus Kriel on 29 July 1997.[6] I was privileged to attend the 1997 re-opening. Seeing the understandable pride of those who had been involved in the planning, design and construction, I could only agree with a paragraph in the information leaflet which had been handed out: 'Another chapter has been written in the history of the Outeniqua Pass and also in the history of road design and construction in our country.'

CHAPTER THIRTY

ANENOUS PASS

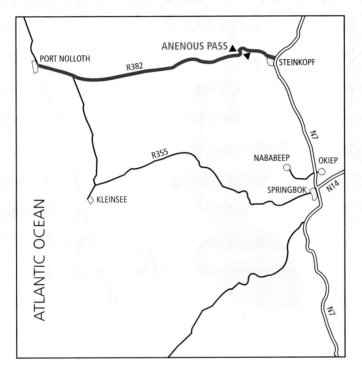

The North Road from Cape Town to Namibia runs through Springbok in Namaqualand. About 50 kilometres north of that town and 75 kilometres south of the Orange River, a surfaced road takes off to the west, past Steinkopf, on its way to Port Nolloth and Alexander Bay. About 20 kilometres along this road, Anenous Pass will take you down the escarpment, from the rocky Hardeveld, where all the mineral deposits have been found, to the 70-kilometre-wide 'beach' of the Sandveld, where diamonds are found. The escarpment forms a sharp, steep and definite barrier, and it is no surprise to find that Anenous is derived from the Khoikhoi *nani=nus*, meaning 'the side of the mountain'.

THE OLD PASS

The first commercial use of a road pass at Anenous was made by the 'copper riders' transporting copper ore from the mines around Okiep and Concordia for shipment overseas from Port Nolloth. Although the major shipping port was Hondeklip Bay there was always a certain amount going via Port Nolloth. In Namaqualand, vegetation grows where the rain has fallen, and on occasions there would be more fodder along the Port Nolloth route than on the way to Hondeklip Bay.

The Okiep–Port Nolloth narrow-gauge railway (NLSA 9487)

Between 1869 and 1876 Richard T. Hall pushed the narrow-gauge railway through from the port to the mining area for the Cape Copper Company. After that, all the ore outbound and almost all the freight inbound were carried by the railway until it was closed down in 1941. But road transport in Namaqualand had been growing for ten years or so before that, and the Port Nolloth road was regaining its importance. When I went to Namaqualand in 1949, the existing Anenous Pass had obviously had its surface and drainage cared for, but I doubt whether its location had altered much over the years: it ran down an obvious nose to reach the sands. This very fact meant that if the grade was to be improved, an entirely new location would have to be found. Nothing more could be done to improve the existing pass.

THE 1950 LOCATION

This, then, was the situation when John Williamson appeared unannounced at my 8x8 wood-and-iron office on my Provincial construction unit, demanding a supply of flags on two-metre poles. He swept me off with him the next morning to do a location reconnaisance for a new Anenous Pass – not that I needed much sweeping, mind you.

John was a *padmaker* imbued with tremendous energy, backed by an ability and knowledge which he expanded throughout his career. When he left the Cape Provincial Roads Department he went to what was then the South West Africa Roads Department as the Chief Engineer. During his years of service there, he left his mark indelibly on the territory that is now Namibia. He was a *padmaker* who will long be remembered and whose works will still be with us when he himself is no longer a memory.

John never carried water when walking in the veld. His practice was to start drinking water as soon as he woke up and to continue forcing as much of it down as he could while he washed, shaved, dressed, and drove to the site. His theory was that this would (probably) keep you going, and I must say that it seemed to work (most of the time) although it is quite against the current theory and practice in organised long-distance running, walking and cycling. Of course, when it didn't work (as on that day) the result was rather unpleasant.

We set off from Springbok with John's wife so that we arrived at the top of the old Anenous Pass at sunrise. We studied the mountainside through our binoculars. We pranced along the rim like mountain goats and looked down from the north, then retraced our steps and looked down from the south. We then drove down the pass and looked at the mountain from the bottom, and we thought there might just be two possible routes. So we drove up the hill again and, while Mrs Williamson took the car down to the bottom and off on a sand track to a position which we had indicated to her, John and I started traversing down the mountainside with Abneys and flags. Halfway down we found that the first route would not 'go', and so we climbed up again and tried the second route. This one we

The old Anenous Pass (Cape Archives AG 14380)

were able to take down to the bottom of the mountain all right, but it was the sort of route that would gladden the heart of anybody who had a monopoly on the sale of explosives: very tough, very hard and very expensive.

By now it was well after midday in the middle of the summer in Namaqualand and we were rather tired and hot. I, for one, was feeling thirsty. Mrs Williamson was dutifully sitting in the shade alongside the car where we had thought the first traverse would come down, but we were quite a few kilometres to the north of that position, so there was nothing for it but to walk along the foot of the mountain back to where the car was.

The sand was soft. I have a standard three-foot stride but John's was an inch or so longer than mine. We started off abreast, strung about with Abney levels and binoculars and carrying flags, but after a while I found that I couldn't manage that extra stretch and I started dropping behind. By this time we were both a bit staggery. Every ten minutes or so John would do a sort of sideways shuffle and look back over his shoulder to make sure I was still on my feet.

Eventually the car came in sight. Mrs Williamson, sitting in the shade, saw us coming and gave us a cheerful wave. However, she had been told to wait in that place, and so she did not drive along to meet us. John arrived at the car maybe a quarter of an hour later, and I was some time behind him. When we had recovered and had soaked up some much-needed liquid, he said that he had not dared to wait for me because if he had stopped he didn't think he would have got going again. It was quite a day.

To expand our field of investigation we arranged for the loan of the local mine inspector's open Jeep. Some days later we found ourselves jeeping along the track where Thomas Hall's old narrow-gauge railway line used to run. For some reason all the culverts had been removed, leaving vertical-sided ditches across the formation. Crossing these called for considerable ingenuity, and when we got to the steeper side-slope sections we loaded up some old sleepers and used them as rather precarious and creaky bridges.

We finally worked our way around to the western face of the escarpment and to a point where we were able to see an attractive road route branching away from the railway route at a steeper but eminently acceptable road grade.

We now had to get down from the mountain, and I assumed in my ignorance that we would continue along the rail track, bridging the culverts as we went, however tedious that might have been. However, John looked down the mountainside, which I swear was steeper than the natural angle of repose of the material, grinned, did a series of nifty backing-and-filling manoeuvres that got the Jeep facing directly downhill, and said, 'Right! Let's go.'

I held on to all the odd bits of the Jeep that I could get my hands or feet to and we set off straight down that scree slope, accompanied by a cloud of dust, stones and small rocks. It was a wonderful experience, but not one that I would willingly repeat. In hindsight, we were probably lucky the Jeep kept going straight, but John did it, and I must say that it was very nice not to have to walk back to the car through the soft sand when we got to the bottom.

THE 1953 PASS

This pass route was developed and found to be so superior to the alternatives that the roads both to east and west were relocated to fit in with it. This involved the construction of 16 kilometres of new road, of which five kilometres could be regarded as being the pass proper. Here the fact that the old railway trace was available for use as the pioneering construction track greatly assisted in getting the job started. The work was carried out at a cost of £26,000, and was of sufficient importance to the region to justify the Minister of Lands, P.O. Sauer, coming up to perform the opening ceremony on 10 July 1953.[1]

Climbing the escarpment had always been the critical section on the Port Nolloth–Springbok road, and the provision of this new Anenous Pass, with its easier gradient and improved geometrics, greatly eased the road transport of goods from the port. But it still had a gravel surface – all that was justified at that time.

As time passed, the volume of traffic increased. The road system throughout the Republic was slowly but surely improved to cater for this growth. In Namaqualand the surfacing of the road between Springbok and Steinkopf had been completed in 1962, and the black top had been extended to the South West African border at Vioolsdrift by 1966. In 1967 some 13 kilometres of the road from Port Nolloth towards Steinkopf had been permanently surfaced, up to Gemsbokvlakte, where the high-standard gravel road took off to the south around the coastal diamond area, on its way to the diamond mines at Kleinsee.

THE 1979 PASS

So in due course the surfacing of the road from Steinkopf via Anenous Pass to Gemsbokvlakte was necessary to close the system. The road had to be surveyed, designed and built. One of Pieter Baartman's survey memories is of 'Anenous Pass, where the broken bottles made continuous glittering ribbons along the road shoulders, and where we came across a dump of crayfish shells whose stench was noticeable from 100 metres away and where we saw the largest flies ever. Jannie van der Westhuizen, then Roads Inspector, remarked that they were so large that "hulle sit 'n half kroon sommer toe."'[2]

The design was done by Jeffares & Green and the construction by Savage & Lovemore. Edward Sunde supervised the contract, and has provided a monograph covering some of his recollections of this interesting project.[3]

The 1979 Anenous Pass (Graham Ross)

'The reconstruction followed the existing alignment fairly closely and involved the widening of all the rock cuts and two or three large new cuttings to improve the horizontal alignment. The rock was predominantly granite with a tendency to break into large blocks when blasted.

'In the pass section there was no possibility of the traffic being able to bypass the construction except on the short sections of new alignment. The traffic count is fortunately low on this section, so by limiting the size of the blasts and restricting them to a fixed time each day, it was possible to carry out the work under traffic although in some cases the riding surface was very rough. The steep grades and the nature of the material made it virtually impossible to move the material uphill, so the work was planned in such a way that nearly all the shot rock material could be moved downhill using Cat 35-ton dump trucks and a Komatsu 355 45-ton dozer to spread and place the rock fill.

'There is of course no surface water in this area and underground water is difficult to locate, but one reasonably strong borehole was drilled approximately five kilometres from the top of the pass and this was pumped to a temporary storage dam at the top of the pass. A gravity pipeline was utilised down the pass to the lower plateau but it was necessary to construct a break-pressure dam halfway down to prevent the pressure from bursting the pipe. The salts in this water made it unfit for human consumption but, owing to the heat, new employees invariably disregarded the warnings and found, to their cost, that the water was more efficient than any Epsom Salts they could purchase.

'Unfortunately we did not come across any diamonds and there were no suggestions that any of our quarries were in diamond-bearing material!'

The contract was completed in 1979, and the cost was R8,342,948.[4]

Today Anenous Pass is a very pleasant drive. There is a viewpoint on the neck about halfway down, and from there you get an excellent view of the forbidding escarpment, stretching away to the south. When one remembers that it extends for 150 kilometres or so one appreciates just what a barrier to travel it constituted. Looking at those rock faces, one no longer wonders why Colonel Robert Gordon, going from the Orange River mouth to Springbok in 1779, travelled via Hermanus Engelbrecht's farm Ellemboogfontein, southwest of Kamieskroon. When Gordon found it impossible to take his wagons east along the riverbed, it was so obvious to him that he had to go south for 250 kilometres to round the southern tip of the escarpment before heading north again to Springbok that he does not even mention the reason in his records.[5]

If the sun is right, you will also be able to pick out the scar of the old Anenous Pass, and appreciate how steep it is and just why that line could not be upgraded.

CHAPTER THIRTY-ONE

THE GAMKASKLOOF PASSES

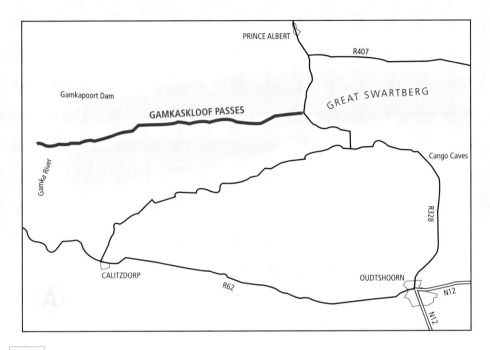

The road to Gamkaskloof takes off from the R328, between Oudtshoorn and Prince Albert, two and a half kilometres north of the summit of the Swartberg Pass. I like to call it the Gamkaskloof Road, but it has also been known as the Otto du Plessis Road, and the Road to Hell, for reasons which will, I hope, become clear later.

The road runs to the westward until it reaches the Gamka River. On the map the crow-flight distance measures 40 kilometres, but the road distance is nearer 57. It is of course a 'slow' road, not only because of the surface and the geometrics, but also because you will want to stop to look at the rugged scenery and to feel the stillness, and probably to wander off into the veld for a while. It is not a through-road. I have driven it a number of times, in my ordinary old car, just for the pleasure of travelling the road.

The Gamka River breaks out of the Swartberg range about ten kilometres northwest of Calitzdorp. It is the only ground-level poort through the mountains between Seweweekspoort and Meiringspoort, but its sides are so steep, its bed on occasions so narrow, its floods so vigorous, and its direction so sinuous that it is not possible to put a road along it. About midway through the mountains the river is crossed by a fertile east–west valley, about 16 kilometres in length. This is the farm Gamka Poort, which was granted to Petrus David Swanepoel in 1841: the survey diagram states that it is 'bounded East, West, North and South by rocks, mountains, inaccessible except along the bed of the river'.[1] A little

The Gamka River at Gamkaskloof (Graham Ross)

community lived and thrived in this valley, happy and contented in their isolation from so-called civilisation.

In his entertaining account of commando warfare in the Anglo-Boer War, Deneys Reitz gives a delightful description of the valley and its hospitable inhabitants. His commando stumbled on the valley when crossing the Swartberg mountains in 1901 to get away from the 'khakis' in the Little Karoo. I shall quote some paragraphs from his fascinating book *On Commando*.

> We could see a long narrow canyon lying at our feet, its sides closed in by perpendicular cliffs. On the floor of the chasm, a thousand feet below, we made out a cluster of huts, and, thinking to find natives there to guide us, we went down in a body to investigate, leaving the horses in a ravine to look after themselves. We climbed through a fissure in the crags, and reached the bottom soon after sunset. As we approached the huts, a shaggy giant in goatskins appeared and spoke to us in strange outlandish Dutch. He was a white man named Cordier, who lived here with his wife and a brood of half-wild children, in complete isolation from the outside world ...
>
> We were received with uncouth but sincere hospitality, and we applied ourselves to the goats' meat, milk, and wild honey that were placed before us. Cordier told us that no

One of the zigzag passes on the Gamkaskloof road (Graham Ross)

British troops had ever penetrated this fastness and that we were the first Boers to do so. He had heard vaguely of the war, but his knowledge of the events of the last two years was scanty.

... in the evening [we] toiled up the cliffs again, accompanied by our host and some of his colts, who stayed with us around our camp-fires, and led us the following morning across rugged mountains, until by dark we looked down at last upon the northern plains ...

We spent another night on the heights, and, parting from our guides at daybreak, climbed down the slopes to level ground and headed across the plains ...

I find it interesting that Cordier led the Boers across the high mountains, and not up the Gamka River bed. I have come to the conclusion that they probably could not get their horses down the 'perpendicular cliffs' to the river level, but there may well have been some other reason.

At this time there were two commonly used routes into the valley, both from the north. True, Gamkaskloofers had on rare occasions made their way down the Gamka River bed to Calitzdorp – where, incidentally, the settlement was known as Gatplaas – but this was apparently only possible when the river was very low. The main routes were along the riverbed to the north, where the Gamkaspoort Dam currently blocks the way, or via a steep, zigzag path on the face of a 450-metre, near-vertical cliff face about five kilometres to the west, connected to Bosluiskloof to the north. Pack-mules could pick their way up the riverbed, but I am informed (I have not been there myself) that 'Die Leer' (The Ladder), as the footpath is known, is suitable only for pedestrians. Apparently Gustaf Nefdt, one of the Kloof farmers, carried a heavy Dover coal stove down the Ladder, lashed to his back. One slip of his foot could have meant the stove falling to be smashed to smithereens on the valley floor – and Gustaf too, of course.[2]

What about the appellation 'Die Hel'? Dr P.C. Luttig, who visited the valley as part of his medical practice, says that the Gamkaskloofers gave the name to a small, solitary, cut-off farm (Boplaas) in a separate gorge adjacent to and west of the main kloof, and that it is here that 'Die Leer' reaches ground.[3] People (other people, not the inhabitants of the valley) love this sort of tag, and it appears as if the name 'Die Hel' came to be applied incorrectly to the whole kloof. The name certainly stuck and was in common usage for many years – in fact, you will find it on the topo map. It was really only after the road investigation brought more contact that the inhabitants' dislike for this appellation was made known, and the name Gamkaskloof came into vogue.

The settlers were apparently happy in their peaceful isolation, satisfied to have only occasional contact with (mainly) Prince Albert, where they would appear with their pack-donkeys carrying dried fruit, beans and peas, and sometimes some wheat, to trade for a few luxuries. But in 1959 requests were received for a road to be provided into the valley. There were also plans under consideration for the construction of what is now the Gamkaspoort Dam at the northern end of the poort.

The Administrator, Dr Otto du Plessis, decided to go and see for himself. I think that this was very brave of him, as he must have known that the trip would entail considerable inconvenience and physical discomfort. Dr Du Plessis, with Rykie van Reenen of *Die Burger*, Ton Vosloo of *Landstem*, an *African Mirror* man with his heavy old camera, and Peter Younghusband of *Panorama*, scrambled down the Gamka River gorge on horseback, riding where possible but leading the horses when the going became tough. Louis Terblanche, the Provincial District Roads Engineer, accompanied by Sampie Luttig, Prince Albert's Mayor and Deputy Chairman of the Divisional Council, and the local Road Inspector, Van Rensburg, went down Die Leer. They were met at the top by Gustaf Nefdt, of Dover stove fame.

Some of Louis Terblanche's recollections about this road follow.[4]

'Each person carried a rucksack with some food and clothes and I had an Abney level to check the gradients with. When we arrived tired and thirsty down in Die Hel at Nefdt's farmhouse, Sampie took a bottle of clear liquid from his rucksack, pulled the cork and gave it to me. As I was dying of thirst, I took two big gulps – and my breath was completely taken away. I had to squat down to recover from the neat witblits!

'We slept that night in the kloof after a big braai and lots of medicine (witblits) and then walked back up the Gamka valley the following day. There were no possibilities for a road either via Die Leer or alongside the Gamka River.

'The Administrator promised that further investigations would be made.

'A week or so later, Prince Albert's Roads Inspector, Van Rensburg, and I drove a little way along a forestry track starting from the Swartberg Pass and walked to Die Hel and back and saw (with the help of my Abney) that a road was possible.

'Province made a sum of about R30,000 available and the Divisional Council used an HD11 Allis Chalmers bulldozer to build the road, which Roads Inspector Van Rensburg and the Provincial Roads Inspector, Riley, from the Oudtshoorn District Roads Engineer's Office pegged out.'

The really heartening and deciding factor was that, while the western approach to Gamkaskloof, from Bosluiskloof, was hard granite which would have needed blasting, Louis found that the eastern approach from the top of Swartberg Pass was over a softer shale, which could be ripped and bulldozed.[5] The Divisional Council tackled the project sensibly, with a small unit consisting of dozer operator and foreman with twelve labourers. They worked away from March 1960 to August 1962, when the road was formally opened by the new Administrator, Dr Nico Malan.

The road consists of 14 kilometres of easy forestry road, then 26 kilometres of up-and-down mountain road which ends spectacularly with a 579-metre drop into the eastern end of the Gamkaskloof valley, along whose bot-

Dr Otto du Plessis rode in on a white horse to see for himself (Louis Terblanche)

tom the road then runs quietly for a further 10 kilometres to the Gamka River causeway.[6]

Burman says that 'the road ... constitutes a mountain pass far tougher than the Swartberg Pass ... it twists and turns, descending one kloof after another, rising in short zigzags up the opposite wall, to repeat the process again'.[7] When I came to list the passes along the route, the only one I felt I could isolate was the final one-in-six 579-metre drop down Eland's Pass. This was something special. Otherwise the passes tended almost to run the one into the other, and I decided they could be lumped together as 'the Gamkaskloof Road'.

◆◆◆◆◆◆◆◆◆◆

So a road had been provided into (and out of) Gamkaskloof, as requested by the local inhabitants. It is worth looking at the effect this has had on a previously isolated but content and close-knit community.

Deneys Reitz mentions only the Cordier family in 1901. When the road was opened in 1962 there were about 120 people in 20 families, living in the Gamkaskloof valley. Burman says that in 1968 the population of the valley was approximately 100 souls, divided into some 17 families.[8] In her sketch plan of the kloof Helena Marincowitz shows fourteen occupied houses in 1993. In April 2002 I picked up a pamphlet advertising self-catering accommodation and camping on a lovingly restored farm. One is invited 'to see old-time farming practised by the only born Gamkaskloofer still living in the valley'.

Louis Terblanche visited the valley in September 2000, for the first time since 1961. His heart-felt comment was this:

'What a disappointment. Whereas in 1960 there had been a flourishing community of more than 20 families with about 120 people in all, 21 children in the primary school, and green fields, peach and apricot orchards, vineyards, etc., now there is only bush and a wilderness with troops of baboons. It made me terribly sad to think that unlike all the other roads I had planned, designed and built through the years, the road to Die Hel had had the opposite effect to that desired and led to destruction and depopulation. If I could have foreseen this, I would definitely not have submitted my recommendation for building the road – I still reproach myself for having done so.'[9]

I feel for Louis. But I think that he blames himself unfairly. Over the past decades there has been a growing movement of people from the farms to the towns, and the young folk would have tended to leave the hard (if pleasant) farming existence in Gamkaskloof for the bright lights of town, whether the road had been there or not. However this may be, the thought does lead me into the final chapter of this book, where I try to do some crystal-ball-gazing, and ask the question: Do mountain passes ever die?

CHAPTER THIRTY-TWO

WHAT OF OUR MOUNTAIN PASSES IN THE FUTURE?

Our mountain passes should never be allowed to deteriorate to this state (World Highways)

DO MOUNTAIN PASSES EVER DIE?

Yes. There are some passes which have been built for a special purpose, such as to give access for the mining of a rich lode in the mountains. When the lode is worked out or, rather, when the working of the lode becomes uneconomic, the operation closes down and, the need having been removed, the pass 'dies'. There are undoubtedly numbers of examples of this: I shall just quote the case of the pass which was built over Du Toit's Kloof in the 1870s to give access to a manganese mine opened there. When the mine proved to be uneconomic, the pass was abandoned.[1]

Some might say that the Mostertshoek toll road 'died' when Michell's Pass was constructed, or that Grey's Pass 'died' when Piekenierskloof Pass was built. But I don't think that this would be fair. I should prefer to say that they were resurrected in a new form to meet the demands for a pass of a higher standard.

Then we come to the Gamkaskloof passes, on an access road which was built to serve a particular purpose the need for which has now gone – the Gamkaskloofers have virtually all departed. But these passes have not died: the justification has just changed. The

Gamkaskloof Road has metamorphosed into a forestry and tourist road, opening up vast wilderness areas for the delight and enjoyment of increasing numbers of people. Similarly, Groot River and Bloukrans Passes and the 1951 Great Brak Pass may have been replaced in their role as carriers of through-traffic by roads and passes of a higher standard, but they are still essential providers of service to the local communities.

To sum up, no mountain passes are built without justification. In some cases the justification may change in character over time. In only a very few cases will all justification disappear and the pass be abandoned. All this appears to me to mean that we must look after our mountain passes in the future.

MAINTENANCE

The initial emphasis should be on a high standard of maintenence. Our mountain passes constitute an important and valuable investment of funds, which must be looked after properly. This requires skilled and experienced staff, preferably familiar with the area and the conditions there, and with the requirements of road maintenance in a pass.

In this connection, I think I detect a tendency in certain quarters to escape their maintenance responsibilities by hiding behind the current trend towards providing more local employment by privatising the work. It is no good farming out the maintenance, and then replying to criticism of the work standard by blaming another party. Of course, privatising roadworks to local people is of itself nothing new – it has been done for many years – but the road authority must retain responsibility for the work.

I think it is self-evident that our mountain passes in the future will require and justify a high standard of maintenance – and the allocation of adequate funds to make this possible.

DESIGN STANDARDS

Design standards of roads change over time, as the vehicles which use the roads themselves change. Hilaire Belloc defined the relationship between road and vehicle thus: 'It was the Vehicle that made the Road. It was the Wheel that made the Vehicle, and the Vehicle having made the Road, the Road reacted back upon that which made it: and though we cannot say that the Highway has, in changing, created a change in Vehicle, yet we can say that, had it not changed, the new Vehicle would not have come into being.'[2] This delightfully convoluted quotation to my mind illustrates the complicated and interactive relationship between the road and the vehicle.

One talks of 'the pace of the ox'. The ox does not trot, nor (unless stampeding in terror) does it gallop. Its pace is relatively unaltered whether pulling on the level or up a (reasonable) gradient. In the ox-wagon days many farmers, living at the foot of mountain passes, kept spare spans of oxen which they hired out to travellers as extra 'ox power' for the ascent, enabling the 'pace of the ox' to remain relatively unchanged. The *padmaker* of that time built his pass at a gradient to suit the needs of these ox-wagon travellers.

When the major change in road vehicles happened in the early 1800s, with the introduction of lighter horse-drawn vehicles capable of higher speeds, the road engineer adjusted the pass gradients to enable the ascent to 'be performed at a brisk trot', as Michell said of Sir Lowry's Pass in 1836, and so that 'a single horse with a gig may trot either up or down the whole length', as Henry Fancourt White said of Great Brak River Heights in 1848. This flattening of the gradient had a beneficial side effect in that ox wagons were able to descend these grades without locking a wheel, or inserting the iniquitous *remskoen*, which ploughed up the road surface.

Came the internal combustion engine, and the increased vehicle size and speed which this advance enabled required greater sight distances for safety (hence larger curve radii), stronger and wider road beds, and still another reduction of critical gradient. And as the volume of traffic increased, the stage was reached on certain road where dual-carriageways and even freeway facilities were required and justified. Thus we see a dual-carriageway freeway built up Great Brak Pass and, where this was not feasible – as with Du Toit's Kloof – a tunnel of considerable length bored through the heart of the mountain to provide the required additional capacity and safety.

This all leads up to the truism that our mountain passes of the future will require improved geometric standards and increased capacity. (Especially is this the case in our country where, as in many others, rail transport is fighting a losing battle, struggling to retain for itself a modicum of freight traffic.)

NEW FACILITIES

Our population is growing; at the same time, unemployment is unacceptably high. We must expand our economy to provide employment, and generally to meet our country's demands for growth.

There are very few commercial or industrial undertakings which can function without road transport. So not only must the existing road facilities be adequately maintained, but new roads and streets must be provided to serve our burgeoning economic development.

Forgive me if I appear to have got away from the question of mountain passes. I haven't really. One needs roads to get to the passes. And just as 'no man is an island ... every man is a piece of the Continent, a part of the main', so no road system or network can function adequately if one or some of its links are missing.

Our mountain passes are so often essential and irreplaceable links in the network. In a city one can generally use a parallel street if the street which one wished to use is being dug up or has been blocked by an accident, but with a mountain pass this is not so. On 2 August 1995, when we were on our way to the Kalahari Gemsbok Park, we sat at the foot of Van Rhyn's Pass for half a day with our caravan while people struggled to sort out a semi-trailer combination which had its horse dangling over the edge of the pass. In mid-afternoon a senior police officer came back and told us that there was no way they were going to open the pass that day – or early next day, from what he could see. So instead of going via Calvinia and Keimoes we turned round and went via Springbok and Kakamas. Luckily for us, we had no business in the Hantam – others were not so fortunate.

To get back to my crystal ball, this means that where the need arises or economic development demands it, we must be prepared to provide the required mountain passes in the future.

The point is, of course, that our mountain passes, being such critical links in our road networks, must have extra-special maintenance care, and must be upgraded as necessary to meet the demands of our growing economy. And to do this we need competent, experienced, and caring *padmakers*, and the funds to enable them to do what is necessary.

Without such adequately funded care our mountain passes will be unable to support the essential socio-economic development of our country. Someone must see to it!

APPENDIX

LIST OF CAPE MOUNTAIN PASSES

This Appendix lists about 490 separate road mountain passes in the Cape of which mention has been found during research carried out between 1993 and 2002. The list also constitutes Section A of *Mountain Passes, Roads and Transportation in the Cape: A Research Document*, prepared by the author. It consists of the following Sections:

A: Index of Cape mountain passes
B: Schedule of construction completion dates
C: Chronology of passes, roads and transportation
D: Annotated bibliography

The various chapters which form the main part of this book have been written using some of the information which is contained in the entries in the Chronology of the Research Document.

Should further detail be desired, you are referred to the Research Document itself. Copies of the second edition have been deposited in Western Cape Provincial Department of Transport; African Studies Library, University of Cape Town; Africana Library, Kimberley; National Library of South Africa: Cape Town Division; Bartolomeu Dias Museum, Mossel Bay; C.P. Nel Museum, Oudtshoorn; and the Fransie Pienaar Museum, Prince Albert. Copies are also held personally by a number of interested people. If you would like to know where the copy nearest to you is located, please contact the author care of David Philip Publishers.

INDEX OF CAPE MOUNTAIN PASSES

NOTES

1. The term 'mountain pass' as used in this work is taken to include poorts that give access from one side of a mountain to another, e.g. Meiringspoort. And passes do not necessarily have to go up, over a mountain and then down again; they can also go down to a river and then up the other side, e.g. Homtini Gorge; or they can just go down (or up!), e.g.: Hex River Pass.
2. The annotation 'aka' (also known as) is followed by alternative spelling, or other names for the pass.
3. An annotation 'See also' is followed by the names of other passes along the same route.
4. The dates indicate where the Chronology (in Section C of the Research Document) may be entered for further information about a particular pass.

Aasvoëlkrans: off R63 south of Nieu-Bethesda.
Adolphus Poort: off R75 southeast of Wolwefontein.
Addo Height: possibly Suurberg Pass. 1858
Agter Witzenberg Pass: on DR1468; off R303 at the top of Gydo Pass (earlier pass known as Schurfdeberg Pass; see also Witsenberg Pass). 1780
Akkedisberg Pass: on R326 northeast of Stanford (aka Aagedisberg Pass; Kleijne River Pass; Clyne River Cloof). 1776
All Saints' Nek: on R61 east of Engcobo.
Allemanshoekpas: on DR2254, between Merweville and Sutherland. 1994
Allemans Poort: on N6 south of Jamestown. 1953
Amandelhoogte: on N9 northeast of Graaff-Reinet (see also Goliatshoogte, Lootsberg Pass, Naudesberg Pass, Perrieshoogte). 1858
Anenous Pass: on R382 west of Steinkopf (aka Aninaus Pass). 1953
Anysberg Pass: off R323 southwest of Laingsburg.
Appelfontein Hoogte: on R354 north of Matjiesfontein (see also Rooikloof, Verlatenkloof).
Attaquas Kloof Pass: early pass north of Mossel Bay (aka Lange Clouw; Lange Cloov; Long Kloof; Attaqua Kloof; Attaquass Kloof; Artaquas-Kloof). 1689

Baillie's Pass: off N7 northeast of Kamieskroon (aka Bailey's Pass. See also Klein Nourivierpas). 1863
Bain's Kloof Pass: on R303 east of Wellington. 1853
Banghoek Pass, Stellenbosch (aka Banhoek Pass; later Helshoogte Pass). 1705
Barberskrantspas: on N10 south of Cradock.

199

Barkly Pass: on R58 north of Elliot. 1966
Baster Voetpad: on R393, mountain road northeast of Elliot. 1862
Battle Neck Pass: see Bottelnek Pass.
Baviaanskloof: R332, Willowmore to Patensie (see also Groot River Pass; Nuwekloof Pass). 1886
Beletskloof: off R61 west of Cradock.
Bell River Pass: on R396 northwest of Maclear (see also Rebelhoogte, Naudesnek; Pot River Pass). 1896, 1911
Benjaminshoogte: on R58, 20 km southeast of Lady Grey.
Bergkloof: see Cloete's Pass. 1850
Bergnaarspadpas: off R383 southwest of Postmasburg.
(Die) Bergpas: on R313 north of Prieska.
Bidouw Pass: off R364 north of Wuppertal (see also Hoekseberg, Kleinhoog, Koudeberg and Uitkyk Passes).
Biesiespoort: off N12 south of Victoria West.
Biesiespoort: N14 to R64 north of Kakamas.
Blaauwkrantzpas: on R67: see Bloukrans Pass.
(Die) Blespas: off R67 northwest of Fort Beaufort (see also Blinkwater Pass).
Blinkwater Pass: off R67 northwest of Fort Beaufort (see also Die Blespas).
Blinkbergpas: off N7 southeast of Clanwilliam (see also Cedarberg Pass, Grootrivierhoogte, Kriedouw Kloof, Nieuwoudtspas and Varkkloof)
Blomfontein Pass: off R390 south of Cradock (aka Bloemfonteinberg Pass)
Blouhoogte: on R61 east of Cradock (see also Plankfontein Heights)
Bloukrans Pass: on R102 east of Plettenberg Bay (aka Blaauwkrans; Blaauwkrants, Blue Krantz. See also Groot River Pass). 1883
Bloukrans Pass: on R67 southeast of Grahamstown (see also Lushington Pass).
Bloukrans Pass: on R355 south of Calvinia.
Boegoebergpas: somewhere 28/29°S, 22/23°E; probably east of Groblershoop.
Boesmanshoekpas: on R397 northwest of Sterkstroom.
Boesmanskloof: McGregor to Greyton: never completed. 1936
Boesmanspoort: see Perdepoort.
Bokkrans Pass: on DR1523.
Boma Pass: off R352 southwest of Keiskammahoek. 1850
Bongolonek: on R392 northeast of Queenstown (aka Bonkolo Nek).
Bosluiskloofpas: on DR1720 northeast of Seweweekspoort (aka Boschluis Kloof). 1863

Bothas Post: off R67 south of Fort Beaufort.
Bothmaskloof Pass: on R46 southwest of Riebeek Kasteel (aka Botmans Kloof). 1700s
Bottelgat: off R344 north of Adelaide.
Bottelnek Pass: off R58 southeast of Barkly East (aka Battlenek Pass)
Botterkloof: on R364 northeast of Clanwilliam (see also Pakhuis Pass, Klipfontein Pass). 1877
Braambos Pass: on R344 north of Adelaide (see also Bushnek Pass)
Brakdam se Hoogte: on N7 south of Springbok (aka Garies Tweede Hoogte. See also Garies Hoogte and Burke's Pass). 1970
Brakkloof: off R67 northeast of Grahamstown.
Brakkloof: R317 to R319 north of Bredasdorp.
Brakpoort: off N12 east of Victoria West.
Brak River Heights Pass: north of Great Brak Village (aka Brakhoogte; Great Brak River Heights; the 'Old Heights'). 1848
Brand se Berg Pass: on DR2198.
Brandwaghoogte: on R328 north of Mossel Bay (see also Robinson Pass). 1869
Bronkhorst Hoogte: on DR1463.
Brooke's Nek: on N2 south of Kokstad (aka Brooks Nek; Brook's Pass). 1960
Brown's Pass: southeast of Sendelingsdrif, Richtersveld. 1970
Bruintjies Hoogte: on R63 southeast of Pearston. 1965
Buffalo Neck: off N2 northwest of Mount Frere.
Buffalo Pass: off R72 west of East London. 1940s
Buffelsbergpas: southeast of Citrusdal: see Buffelshoek Pass. 1994
Buffelshoekkloof: on R337: see Buffelshoeknek. 1824
Buffelshoeknek: on R337 northeast of Pearston (aka Buffelshoek Pass, Buffelhoek Pass, Buffelshoekkloof. See also Swaerhoekpas). 1824
Buffelshoek Pass: on R337: see Buffelshoeknek. 1824
Buffelshoek Pass: on R303 southeast of Citrusdal (aka Buffelsbergpas. See also Middelberg Pass). 1994
Buffelskloof Pass: from R328 towards Calitzdorp. 1994
Buffelspoort: south of Laingsburg (aka Buffelsrivierpoort, Die Poort). 1960s
Burger's Pass: on R318 northwest of Montagu (aka Koo Pass originally. See also Rooihoogtepas). 1877
Burgers Pass: east of Lady Grey. See Jouberts Pass. 1991
Burke's Pass: on N7 south of Springbok (see also Garies Hoogte, Brakdam se Hoogte). 1970
Bushnek Pass: on R344 north of Adelaide (aka

Bosnek. See also Braambos Pass).
Buyspoort: on N9 south west of Willowmore (aka Buispoort. See also Ghwarriepoort). 1963

Cala Pass: on R393 northeast of Cala.
Caledon Kloof: off R62 southwest of Calitzdorp (aka Welgevonden; Rooielsboskloof; Verkeerde Kloof). 1807
Campherskloof: off R328 south of Klipplaat.
Camphoor Poort: between Langkloof and Oudtshoorn. 1843
Candauw: see Hex River Pass. 1860s
Cango Poort: see Schoemanspoort. 1862
Cardouw Pass: see Nardouw Pass. 1801
Cardouw's Kloof Pass: see Kardouw's Kloof Pass. 1773
Carlton Heights: on N10 north of Middelburg (aka Winterhoek Pass?).
Cats Pad (previously Olifants Pad; later Franschhoek Pass). 1819
Cats Pass: off N2 southeast of Butterworth en route to Mazeppa Bay.
Cedarberg Pass: on DR1487 off N7 southeast of Clanwilliam (aka Sederberg Pass, Cederbergen Pass; previously known as Uitkyk Pass. See also Blinkbergpas, Grootrivierhoogte, Kriedouw Kloof Pass, Nieuwoudtspas and Varkkloof).
Chapman's Peak Drive: on M6 south of Hout Bay. 1922
Charlie's Pass: the road to Morceaux. 1994
Cloete's Pass: on R327 north of Herbertsdale (aka Bergkloof; Cloeteskraal Pass. See also Du Plessis Pass). 1850
Clyne River Cloof: see Akkedisberg Pass. 1776
Cockscomb Pass: on DR1831 off R75 southeast of Wolwefontein.
Coetzeespoort: off R62 northeast of Calitzdorp.
Cogmans Kloof: on R62 southwest of Montagu (aka Cochmans Cloof; Coggelmans Kloof; Coghemans Cloof; Cogmans Poort; Kochmans; Kochemans; Koekemans; Kogmans; Kokmans Kloof). 1877
Colonanek: off N2 north of Mount Frere.
Constantia Nek: on M63 northeast of Hout Bay (aka Cloof Pas). 1666
Cradock Hoogte: on N10 north of Cradock.
Cradock Pass: early pass, north of George (aka Cradock Kloof; Cradock's Kloof Pass). 1812

Daggaboersnek: on N10 30 km north of Cookhouse. 1850
Dasklip Pass: off R365 northeast of Porterville. 1948
Dassiehoek Pass: off R60 north of Robertson (aka Dassieshoek).
De Beerspas: off R344 north of Adelaide, on MR645.

De Bruin's Poort. 1843
De Jagerspas: off N1 north of Beaufort West; known as Wagenaar's Kloof until 1899. 1880
Derdepoort: off R369 north of Colesberg.
Devil's Bellows Neck: on R351 north of Balfour (see also Katberg Pass).
De Waalskloof: off R351 north of Balfour.
Diepkloof Pass: east of Burgersdorp (see also Kapokkraalhoogte, Witkophoogte).
Diep River Pass: on The Passes Road, east of George (aka Klein Keur River Pass; Swart River Pass. See also Homtini, Hoogekraal, Kaaimans, Karatara, Phantom, Swart River, Touws River Passes). 1882
Die Bergpas: see (Die) Bergpas.
Die Venster: see (Die) Venster. 1887
Domerog Pass: east of Hell's Gate in Richtersveld.
Dontsa Pass: on R352 southwest of Stutterheim. 1857
Doringnek: on R335 north of Addo (see also Suurberg Pass). 1858
Doringhoek Pass: see Groot Doringhoek Pass.
Doringrivier Pass: early pass north of Van Rhyns Pass. 1880
Droëkloof: on R29 north of Meiringspoort. 1960s
Droëvoetspoort: on R356 southwest of Fraserburg.
Duiwelskop Pass: early pass, northeast of George (aka Duyvil's Kop Pass; Devil's Kope Pass; De Duivels Kop Pass; Devil's Head Pass; Nannidouw). 1772
Duiwelsnek: on R359 east of Kakamas.
Du Plessis Pass: on R327 south of Herbertsdale (see also Cloete's Pass). 1850
Du Toit's Kloof Pass: on N1 east of Paarl (aka Jan de Toi's Kloof; De Toies Kloof. Earlier: Elephant Route; The Hawequa Cattle Path). 1738
Dwarskloof: off R406 south of Greyton.

Ecca Pass: on R67 northeast of Grahamstown (see also Koonap Heights). 1837–1845
Eerstepoort: on R369 northwest of Colesberg.
Elandsberg Pass: on N6 south of Aliwal North. 1957
Elands Kloof Pass: early pass southeast of Citrusdal. 1798
Elandskloof: off N9 northeast of Graaff-Reinet.
Elandskloof: off R398 about 50 km east of Richmond.
Eland's Pad: see Hottentots Holland Kloof. 1662
Elandspas: off R328, on Gamkas Kloof Road (see also Gamka's Kloof Road). 1962
Elandspoort: on N12 north of Victoria West.
Elephant Route: see Du Toits Kloof. 1949
Erasmuskloof: off N9 east of Graaff-Reinet.

Erekroonspoort: off R329 east of Steytlerville.
Ertjiesvleipas: north of Heidelberg on approach to Gysmanshoek. 1863
Eselsgalpas: see Rooihoogte Pass. 1877
Eseljagpoort: off N9 east of Herold.
Esterhuizen Pass: see Piet Esterhuysen Pass. 1951
Ettrick Hills: on R350 northwest of Grahamstown (see also Hells Poort)

Fetcani Pass: on R393 north of Elliot.
Fincham's Nek: off N6 south of Queenstown.
Fish River Heights: old pass 30 miles north of Grahamstown. 1850
Floorshoogte: see Theewaterskloof.
Floriskloof: on R361 north of Carnarvon.
Fonteinskloof: on R72 west of Alexandria.
Franschhoek Pass: on R45 east of Franschhoek (aka Oliphants Pad; Cats Pad; French Hoek Pass). 1819
Fuller's Hoek Pass: off R67 northwest of Fort Beaufort.

Gamka's Kloof Road: mountain road off R328, Swartberg Pass (aka Otto du Plessis Road; road to Die Hel; includes 579-metre-high Elandspas). 1962
Gannagapas: on DR2250, off R354 northwest of Sutherland.
Gantouw Pass: early pass. See Hottentots Holland Kloof. 1662
Garcia's Pass: on R323 north of Riversdale (see also Langkloof, Muiskraalpas, Voetpadskloof). 1877
Garden of Eden Pass: northeast of Knysna.
Garies Hoogte: on N7 just north of Garies (see also Brakdam se Hoogte, Burke's Pass). 1970
George and Knysna (Old) Road: see (The) Passes Road. 1882
Ghwarriepoort: on N9 southwest of Willowmore (aka Gwarriespoort; earlier called Suurberg se Loop. See also Buyspoort). 1963
Gidouw Pass: see Gydo Pass. 1848
Gifberg Pass: off R27 south of Vanrhynsdorp.
Glen Avon Pass: on DR2462.
Goliatshoogte: on N9 northeast of Graaff-Reinet (aka Goliatskraal se Hoogte; Goliadskraal. See also Amandelhoogte, Lootsberg Pass, Naudesberg Pass, Perrieshoogte). 1858
Gordon's Bay–Steenbras Dam Mountain Pass: off R44. 1937
Gqutyini Pass: off R61 northwest of Engcobo.
Great Brak Pass: on N2 east of Mossel Bay (aka Great Brak Heights). 1951
Great Brak River Height: see Brak River Heights Pass. 1848

Great Kei Pass: see Groot-Keirivierpas. 1955
(The) Great Zwarte Berg Pass: see Swartberg Pass. 1888
Greylingspas: on R396 southwest of Barkly East (see also Killian's Pass, Perdenek, Rossouwsberg Pass and Swartnek).
Grey's Pass: early pass, on N7 south of Citrusdal (aka Gray's Pass; as Piquinierskloof before 1858; and as Piekenierskloof after 1958). 1858
Groot-Doringhoekpas: on R391 northeast of Hofmeyr, on MR663.
Groot-Keirivierpas: on N2 southwest of Butterworth (aka Groot-Keipas, Great Kei Pass, (The) Kei Cuttings). 1955
Grootrivierhoogte: off N7 southeast of Clanwilliam (see also Blinkbergpas, Cedarberg Pass, Kriedouw Kloof, Nieuwoudtspas and Varkkloof).
Groot River Pass: on R102 east of Plettenberg Bay (aka Grootrivierpas. See also Bloukrans Pass). 1882
Groot River Pass, at eastern end of Baviaanskloof (see also Nuwekloof Pass).
Grootvlei Pass: off N7 west of Kamieskroon (see also William's Pass, Killian's Pass).
Grootvleipas: on R58 south of Barkly East. 1966
Gwarina Heights: on N9 southwest of Uniondale (see also Potjiesbergpas). 1962
Gwarriespoort: see Ghwarriepoort. 1963
Gydo Pass: on R303 north of Ceres (aka Gidouw Pass, Gydow Pass). 1848
Gysmanshoek Pass: early pass, north of Heidelberg (aka Hudson's Pass; traverses Plattekloof). 1841, 1860, 1772

Halfmens Pass: southeast of Sendelingsdrif, in Richtersveld.
Hall's Hill: on R346 southeast of King William's Town.
'The Hawequa Cattle Path' (later Du Toits Kloof Pass). 1738
Hells Poort Pass: on R350 northwest of Grahamstown (aka Helspoort. See also Ettrick Hills).
Helshoogte Pass: on R310 east of Stellenbosch (a deviation of the older Banghoek Pass). 1854
Die Hel: see Gamka's Kloof. 1962
Heroldsbaaihoogte: off N2, on original Herolds Bay road. 1911
Heuningklipkloof: early pass northeast of Albertinia. 1720s
Heuningneskloof: on N12 south of Kimberley.
Hex River Pass: on N1 near De Doorns (two passes: the western poort and the eastern climb out of the valley. Aka Hex River Poort; Hex

River Kloof; Hexrivierpas; Heks River Pass; Candauw). 1860s

Highlands Nek: early pass west of Grahamstown. 1958

Hoekseberg Pass: off R364 north of Wuppertal (see also Bidouw, Kleinhoog, Koudeberg and Uitkyk Passes).

Hoekwil Pass: between the Lakes Road and The Passes Road. 1994

Hogsback Pass: on R345 north of Alice. 1932

Homtini Pass: on The Passes Road (aka Barrington Pass. See also Diep, Hoogekraal, Kaaimans, Karatara, Phantom, Swart River, Touws River Passes). 1882

Hoogekraal Pass: on The Passes Road, east of George (see also Diep, Homtini, Kaaimans, Karatara, Phantom, Swart River, Touws River Passes). 1871

Hottentots Holland Kloof: early pass, east of Somerset West (aka Eland's Pad; Elands Path; Elands Pass; Gandou; Gantouw; Hottentots Holland Cloev; T'kanna Ouwe). 1662

Hottentotskloof: on R46 northeast of Ceres (see also Theronsberg Pass, Karoopoort).

Houtbaainek: on Victoria Road (M6), above Llandudno. 1888

Houw Hoek Pass: on N2 west of Caledon (aka Hout Hoeck; Hout Hoek; Hou Hoek; Howe-Hook; How Hoek Pass; original name Cole's Pass or Poespas, i.e. Higgeldy Piggeldy Pass). 1682

Howison's Poort: on N2 southwest of Grahamstown (aka Howieson's Poort). 1850

Hudson's Pass: an improved Gysmanshoek Pass/Plattekloof. 1860

Huisrivier Pass: on R62 west of (aka Huis River Pass). 1896

Ibisi Pass: on R56 northeast of Kokstad (aka Ibisi Heights). 1960

Indwe Poort: on R359 southeast of Indwe.

Jan de Toi's Kloof: see Du Toits Kloof. 1949

Jangora Pass: on R355 northwest of Calvinia.

Janspoorthoogte: on R58 northwest of Burgersdorp.

Jashoogte Pass: on R388 northwest of Graaff-Reinet, on MR606.

Jonkersnek: off R63 southwest of Murraysburg.

Jouberts Pass: off R58 east of Lady Grey (aka Burgers Pass). 1991

Kaaimans River Pass: on The Passes Road (the pass also crosses the Silver River. See also Diep, Karatara, Hoogekraal, Homtini, Phantom, Swart River, Touws River Passes). 1869

Kaaimans Gat Pass: on N2 east of George (aka Quaiman's; Kaymans; Caymmans; Kaaimansgat Drift; Keerom's River; Keerom River – Turnabout River; Kujman's Kloof; Wilderness Pass). 1752

Kaapsepoortjie: on N9 southwest of Aberdeen.

Kafferskraal Pass: off R356 southwest of Victoria West.

Kalk Rand Pass: southwest of Sidbury. 1820

Kamiesberg Pass: off N7 east of Kamieskroon (aka Kamieskroonberg Pass).

Kapieteinskloof: off R399 northeast of Sauer.

Kapokkraalhoogte: off N6 west of Jamestown (see also Diepkloof Pass, Witkophoogte).

Kap River Pass: N2 to R67 east of Grahamstown, on MR475.

Karatara Pass: on The Passes Road, east of George (aka Tsao or Witterivier. See also Diep, Hoogekraal, Homtini, Kaaimans, Phantom, Swart and Touws River Passes). 1882

Kardouws Kloof Pass: early pass south of Citrusdal (aka Kartous; Cardow; Cardauw). 1773

Kareebospoort: on R384 northeast of Carnarvon.

Kareedouw Pass: on R402 south of Kareedouw.

Kareedouw Pass: on R407; see Kredouws Pass. 1994

Kareedouwberg Pass: on R407; see Kredouws Pass. 1994

Karnek: on R344 southwest of Sterkstroom.

Karoo Poort: on R355 northeast of Ceres (aka Bokkeveld's Poort; Karoopoort. See also Theronsberg Pass, Hottentotskloof). 1848

Katbakkiespas: off R303 in Kouebokkeveld (see also Skitterykloof; Peerboomskloof).

Katberg Pass: on R351 north of Balfour (aka New Katberg Pass. See also Devil's Bellows Nek). 1864

(The) Kei Cuttings: see Groot-Keirivier Pass. 1955

Keurbooms Heights: on N2 east of Plettenberg Bay.

Kieskammahoek Pass: somewhere 32/33°S, 27/28°E.

Killians Pass: off N7 west of Kamieskroon (see also Grootvlei Pass, William's Pass).

Killians Pass: on R396 northeast of Dordrecht (see also Greylingspas, Perdenek, Rossouwsberg Pass and Swartnek).

Kingo Hills Pass: off R67 north of Grahamstown (aka King Hills Pass).

Kleijne River Pass: see Akkedisberg Pass. 1776

Klein Berg River Kloof: east of Gouda; see Nieuwekloof. 1860

Kleinhoogpas: off R364, just north of Wuppertal (aka Kleinhoogte; possibly aka Koudeberg Pass. See also Bidouw, Hoekseberg, Koudeberg and Uitkyk Passes).

Klein Katberg Pass: see Michell's Pass. 1851

Klein Keur Passs: on The Passes Road; see Diep River. 1882

Klein Langkloof: off N9 and R339 southwest of Avontuur.

Klein Nourivierpas: off N7 northeast of Kamieskroon (see also Bailey's Pass). 1863

Kleinpoort: on R75 southeast of Wolwefontein.

Klein Straat: off N1 west of Touws River (aka De Straat). 1811

Klein Swartberg Pass: on R323 west of Ladismith. 1880

Klipfontein Pass: on R364 northeast of Clanwilliam (see also Pakhuis Pass, Botterkloof Pass). 1877

Klipspringer Pass: off N1, west of Beaufort West (in Karoo National Park). 1992

Kloof Nek: on M62 between Cape Town and Camps Bay. 1848

Knapdaarhoogte: on R391 northwest of Burgersdorp.

Knee Halter's Neck: 1843

Knelpass: off R390 northwest of Steynsburg, on MR674.

Kogmans Kloof: see Cogmans Kloof (aka Cochmans; Coggelmans Kloof; Cogmans Poort; Kochmans; Kochemans; Koekemans; Kokmans Kloof). 1877

Komgha Heights: on N2 southwest of Grahamstown.

Komsbergpas: off R354 south of Sutherland (on DR2243?).

Koo Pass: see Burger's Pass (aka Koo Mountain Pass; Koodoosberg Pass). 1877

Koonap Heights: on R67 north of Grahamstown (adjacent to Koonap Krantz. See also Ecca Pass). 1837–1842

Koringberg Pass: off N7 north of Moorreesburg.

Koudeberg Pass: off R364 north of Wuppertal (possibly aka Kleinhoogpas. See also Bidouw, Hoekseberg, Kleinhoog and Uitkyk Passes).

Koueveld Pass: on MR309 northwest of Seweweekspoort. 1994

Kraaibos Pass: on R365 (DR2196?) north of Piketberg (see also Nardouwsberg Pass).

Kraairivier Pass: on R58 northwest of Barkly East. 1966

Kredouws Pass: on R407 east of Prince Albert (aka Kredouwsberg Pass, Kareedouw Pass, Kareedouwberg Pass, Kriedouw Pass. See also Witkranspoort). 1999

Kreitzberg Pass: somewhere 31/32°S, 19/20°E (on DR2280?).

Kriedouw Kloof Pass: off N7 south of Clanwilliam (see also Blinkbergpas, Cedarberg Pass, Grootrivierhoogte, Nieuwoudtspas and Varkkloof).

Kriga Pass: somewhere 32/33°S, 20/21°E, on DR2270.

Kromhoogte: off N2 north of Herold.

Krom River Pass: on R58, west of Barkly East.

Kruisrivierpoort: northeast of Calitzdorp.

Kruidfontein Hoogte: on R63 south of Graaff-Reinet.

Kwaaimanspas: off R393 south of Cala.

Lange Cloov/Lange Clouw: see Attaquas Kloof. 1689

Langhoogte Pass: on N2, 6 km east of Bot River.

Langhoogte: on N2, approach to Houw Hoek Pass from east.

Langkloof: off N7 northeast of Garies (east of Studers Pass).

Langkloof: on R323 north of Riversdale (see also Garcia's Pass, Muiskraal Pass, Voetpadskloof).

Langkloof Pass: off N14 northwest of Olifantshoek.

Langkloof: north of Mossel Bay: see Attaquas Kloof. 1689

(Die) Langkloof: R62, Avontuur to Joubertina (aka The Long Kloof. See also Lower Long Kloof). 1861

Langkloof: off R62 northeast of Montagu (see also Oubergpas).

Langkloof: off R406 north of Caledon.

Leeukloof Pass: off R63 north of Graaff-Reinet, on MR604 (see also Osfontein Pass, Waterkrans Pass).

Leeukloof Pass: off R407 west of Willowmore.

Leeukloof Pass: off R381 southwest of Victoria West.

Leeukloof Pass: which one is on DR2317?

Leeupoort: off R388 west of Richmond.

Lessingshoogte: on R398 west of Middelburg, on MR610 (see also Witlieshoogte).

(The) Long Kloof: see (Die) Langkloof. 1861

Lootsbergpas: on N9 northeast of Graaff-Reinet (see also Amandelhoogte, Goliatshoogte, Naudesberg Pass, Perrieshoogte). 1858

Loskop Pass: off N9 north of Graaff-Reinet, on MR605 (see also Rubidge Kloof Pass, Slabberts Pass).

Lower Long Kloof: R62, Joubertina to Kareedouw (see also (Die) Langkloof). 1861

Lundin's Nek: on R393 north of Barkly East (aka Lundean's Nek).

Lushington Pass: on R67 southeast of Grahamstown (see also Blaauwkrantz Pass).

Maanhaarspoort: on N12 south of Victoria West.
MacGregor's Pass: old pass north of Citrusdal. c.1910
Mackay's Nek: on R359 northeast of Queenstown (see also Nonesi's Nek). 1967
Malagaskloof: off R396 northwest of Prieska.
Mancazana Pass: on R344 north of Adelaide (aka Mankazana Pass).
Maroegapoort: off R337 northeast of Willowmore (see also Trompetterspoort).
Matjieskloof: off R63 northeast of Williston.
Meidepoort: off R329 northwest of Steytlerville.
Meiringspoort: on R29 north of De Rust (aka Meiring's Poort). 1858
Messelpadpas: off N7 southwest of Springbok (includes Tiger Kloof. See also Wildeperdehoek Pass). 1871
Michell's Pass: on R46 southwest of Ceres. 1848
Michell's Pass: off R67 east of Seymour, towards Hogsback (aka Klein Katberg Pass; sometimes mis-spelt Mitchell). 1851
Michielshoogte: about 15 km west of Nieu-Bethesda.
Middelberg Pass: on R303 southeast of Citrusdal (see also Buffelshoekpas). 1994
Mlenganapas: on R61 east of Umtata.
Modderpoort: see Telemachus Poort. 1955
Molteno Pass: on R381 north of Beaufort West (see also Rose's Berg. These two adjacent passes were apparently sometimes taken as one pass). 1880
Monesi's Nek: see Nonesi's Nek. c. 1880
Montagu Pass: off N9 north of George. 1848
Moordenaarskloof: on DR1800 (see also Suuranys Pass).
Moordenaarsnek: on R56 south of Mount Fletcher.
Moordenaarspoort: off R27 northeast of Calvinia.
Moordenaarspoort: off N12 north of Victoria West.
Moordenaarspoort: old route of R56 east of Middelburg.
Mostertshoek Pass: early pass, southwest of Ceres (aka Mosterd's Hoek Pass; Musteed's Hock. Replaced by Michell's Pass). 1765
Muiskraal Pass: on R323 north of Riversdale (see also Garcia's Pass, Langkloof, Voetpadskloof).
Munniks Pass: on N9 southwest of Graaff-Reinet (aka Munnikspoort). 1960
Murasiekloof: off R63 northwest of Graaff-Reinet.

Nanaga Heights: on N2: see Ncanaha Heights.
Nannidouw: see Duiwelskop Pass. 1772
Nardausberg Pass: on DR2393 (see also Tierkloof).
Nardouw Pass: eastwards off old road north of Clanwilliam (aka Cardouw Pass; Nardouskloof; The Nardouw). 1801
Nardouwsberg Pass: on R365 (on DR2196?) north of Piketberg (see also Kraaibos Pass).
Naudesbergpas: on N9 northeast of Graaff-Reinet (see also Amandelhoogte, Goliatshoogte, Lootsberg Pass, Perrieshoogte). 1858
Naudesnek: on R396 northwest of Maclear (see also Rebelhoogte, Bell River Pass (1896); Pot River Pass). 1911
Ncanaha Heights: on N2 northeast of Port Elizabeth (aka Nanaga Heights).
Nelspoort: off N1 northeast of Beaufort West.
New Katberg Pass: see Katberg Pass. 1864
Nico Malan Pass: on R67 northeast of Seymour. 1967
Niekerkspas: on R335 south of Somerset East (aka Niekerksberg Pass).
Nieuwekloof Pass: on R46 east of Gouda (aka Klein Bergrivier Kloof; Nieuwe Roodezand Kloof; Red Sand Valley; Roodezand Kloof; Roysand Kloof; Tulbagh Kloof; sometimes Nuwekloof). 1860
Nieuwe Roodezand Kloof: see Nieuwekloof Pass. 1860
Nieuwoudt's Pass: on DR1487 off N7 southeast of Clanwilliam (aka Niewoudts Pass. See also Blinkbergpas, Cedarberg Pass, Grootrivier-hoogte, Kriedouw Kloof Pass and Varkkloof).
Nonesi's Nek: on R359 northeast of Queenstown (aka Nounesis Nek, Monesi's Nek. See also Mackay's Nek). c. 1880
Noorspoort: on R329 north of Steytlerville (see also Waaipoort).
Noukloof: southwest of Ladismith (on R52? or R323?).
Nounesis Nek: see Nonesi's Nek. c. 1880
Noupoort: (probably) on N9 at Noupoort.
Nougaspoort: off N1 south of Touws River.
Nuwehoogte: southwest of Robertson: see Stoepelshoogte.
Nuwekloof Pass: on R332 southeast of Willowmore (the western entrance to Baviaanskloof. See also Groot River Pass). 1886
Nuwekloof Pass: on R46 east of Gouda: see Nieuwekloof Pass. 1860

Oberholsters Valley: on R357 southwest of Brandvlei.
Olifantskop Pass: on N10 north of Paterson. 1955

Olifants Pad (later Franschhoek Pass). 1699
Olof Berghspas: off R356 south of Redelinghuys. 1682
Ongeluksnek: on Lesotho border, west of Matatiele.
Op de Tradouw: on R62 west of Barrydale (see also Poortjieskloof, Wildehondskloof).
Osfontein Pass: off R63 north of Graaff-Reinet, on MR604 (see also Waterkrans Pass, Leeukloof Pass).
Otto du Plessis Pass: off R56 north of Ida.
Otto du Plessis Road: see Gamka's Kloof. 1962
Oubank Pass: somewhere 32/33°S, 20/21°E (on DR2277?).
Oubergpas: on R63 northwest of Graaff-Reinet (aka Oudeberg Pass; Oude Berg Pass. See also Rooikranshoogte, Van Ryneveld Pass, Voetpadhoogte). 1829
Oubergpas: off R62 northeast of Montagu (see also Langkloof).
Oubergpas: off R354 west of Sutherland. 1968
Ouberg Pass: which one is on DR1484? And which on DR2255?
Oudeberg Pass: near Vanrhynsdorp.
Oudeberg Pass: see Oubergpas. 1829
Oudekloof: see Roodezand Pass. 1658
Oudekloof: see Oukloof Pass. 1879
Oude Roodezand Pass: see Roodezand Pass. 1658
Ouhoogte: on R340 northwest of Plettenberg Bay.
Ou Kaapse Weg: on M64 north of Fish Hoek (aka Silvermine Pass, Steenberg Pass). 1811
Oukloof Pass: on MR584, off N1 northwest of Beaufort West (aka Oudekloof; Oukloofpoort). 1879
Outeniqua Pass: on N9 northwest of George. 1951

Paarde Berg Pass: east of Stanford. 1825
Paardekop Pass: early pass, northeast of Knysna. 1772
Paardepoort: early pass, north of Montagu Pass (aka Perdepoort).
Paardepoort: old route of R75, past Wolwefontein, northwest of Glenconnor (aka Perdepoort). 1866
Padkloof Pass: off R64 northeast of Groblershoop.
Pakhuispas: on R364 east of Clanwilliam (see also Botterkloof Pass, Klipfontein Pass). 1877
Pampoenpoort: off R384 southeast of Carnarvon.
Papkuilnek: on R355 south of Calvinia.
(The) Passes Road: R355, George to Knysna (aka old George and Knysna Road; Seven Passes Road. See Swart, Kaaimans, Touws, Diep, Hoogekraal and Karatara River Passes, Homtini and Phantom Passes). 1882
Peerboomskloof: off R355 north of Ceres (see also Katbakkiespas; Skitterykloof).
Penhoek Pass: on N6 south of Jamestown. 1892
Perdekloof: off R27 southeast of Calvinia.
Perdenek: on R396 southwest of Barkly East (see also Greylingspas, Killians Pass, Rossouwsberg Pass and Swartnek).
Perdepoort: on N9 north of Willowmore (possibly aka Boesmanspoort).
Perdepoort (northwest of Glenconnor): see Paardepoort. 1866
Perdepoort: north of Montagu Pass: see Paardepoort.
Perrieshoogte: on N9 northeast of Graaff-Reinet (aka Perry's Hoogte. See also Amandelhoogte, Goliatshoogte, Lootsberg Pass, Naudesberg Pass). 1858
Phantom Pass: on DR1613 northwest of Knysna (see also The Passes Road: Diep, Hoogekraal, Homtini Pass, Kaaimans, Karatara, Swart River, Touws River Passes). 1862
Piekenierskloof: on N7 south of Citrusdal (earlier known as Piquinierskloof (1660) and Grey's Pass (1858); aka Piekenaarskloof). 1858
Piekenierskloof: on R354 southeast of Calvinia.
Pienaar's Pass: 4x4 trail in Karoo National Park. 1950
Pienaarspoort: off R329 north of Steytlerville.
Piet Esterhuysen Pass: on C121 off R303 southeast of Citrusdal (aka Piet Esterhuizen Pass; Baliesgat Pad). 1951
Piqinierskloof: early pass, on N7 south of Citrusdal (aka Piekenaarskloof; Pickaneer's Kloof. Later named Grey's Pass (1858) and Piekenierskloof (1957)). 1660
Pitseng Pass: off R56 southwest of Mount Fletcher.
Plankfontein Heights: on R61 east of Cradock (see also Blouhoogte).
Plattekloof Pass: early pass north of Heidelberg (aka De Platte Cloof; Plattakloof; Platteklip Pass. See also Gysmanshoek; Hudson's Pass). 1841, 1860
Pluto's Vale Pass: on DR2039, off R67, northeast of Grahamstown. 1845
(Die) Poort: see Wuppertal–Eselbank Pass. 1800s
Poortjieskloof Pass: on R62 east of Montagu (see also Op de Tradouw, Wildehondskloof).
Poshoogte: on R62 northeast of Barrydale.
Potgieterspoort: northwest of Oudtshoorn.
Potjiesbergpas: on N9 southwest of Uniondale (aka Potjieskloof; Potjiesberg Poort. See also Gwarina Heights). 1962
Pot River Pass: on R396 north of Maclear (see

APPENDIX

also Bell River Pass; Naude's Nek, Rebelhoogte). 1911
Potter's Pass: off R72, 5 km southwest of East London. 1934
Prieskapoort: on R357 southwest of Prieska.
Prince Alfred's Pass: on R339 north of Knysna. 1867

Qacha's Nek: off R56 northwest of Matatiele.
Quaggasfontein Poort: off R63 south of Williston (see also Snyderspoort).

Ramatselisoshek: off R56 northeast of Mount Fletcher.
Rasfonteinpoort: on R61 northeast of Cradock.
Ravelskloof Pass: on R75 north of Jansenville.
Rebelshoogte Pass: on R396, 14 km northeast of Barkly East (see also Bell River Pass, Naude's Nek, Pot River Pass). 1911
Red Hill: on M66 west of Simonstown.
Red Hill Pass: on R352 south of Keiskammahoek.
Red Sand Valley: on R46 east of Gouda: see Nieuwekloof. 1860
Remhoogte: on R80, Ashton towards Swellendam. 1954
Remhoogtepas: on R63 about 30 km southeast of Victoria West.
Renosterpoort: off R390, about 10 km north of Steynsburg.
Rietfonteins Pass: off R61 northwest of Aberdeen, on MR599.
Rietpoort: off R63 northeast of Williston.
Rietpoortnek: off R63 north of Murraysburg, on MR607.
Robinson Pass: on R328 north of Mossel Bay (see also Brandwaghoogte; aka Ruiterbos to the locals). 1869
Roodezand Kloof: see Roodezand Pass and Nieuwekloof Pass. 1860
Roodezand Pass: early pass, west of Tulbagh (aka Roodezand Kloof, Oude Roodezand Pass and Oudekloof). 1658
Rooielsboskloof: see Caledon Kloof. 1807
Rooiberg Pass: on R318 northwest of Montagu: see Rooihoogtepas. 1877
Rooibergpas: off R62 south of Calitzdorp. 1928
Rooihoogte: on R43 northeast of Villiersdorp.
Rooihoogtepas: on R318 northwest of Montagu (aka Eselsgalpas; Thompson's Pass; Rooiberg Pass. See also Burger's Pass). 1877
Rooikloof: on R354 south of Sutherland (see also Appelfontein Hoogte, Verlaten Kloof).
Rooikranshoogte: on R63 southeast of Murraysburg (see also Ouberg Pass, Van Ryneveld Pass, Voetpadhoogte).

Rooinek Pass: on R323 south of Laingsburg. 1994
Rooinek Pass: on DR2415
Rooipoort: on N12 southwest of Britstown.
Rosesbergpas: on R381 north of Beaufort West (aka Rose's Berg; Roses Pass. See also Molteno Pass). 1880
Rossouwsberg Pass: on R396 northeast of Dordrecht (see also Greylingspas, Killian's Pass).
Rossouwspoort: off R337 south of Pearston.
Roysand Kloof: see Nieuwekloof Pass. 1860
Rubidge Kloof Pass: off N9 north of Graaff-Reinet, on MR605 (see also Slabberts Pass, Loskop Pass). 1994
Ruiterbos Pas: early bridle pass north of Mossel Bay (aka Ruitersbosch; this name also applied to Robinson Pass by locals). Mid-1700s

Saaipoort: off R63 southwest of Williston.
Sandhoogte: on R355 west of Springbok, before Spektakelpas (see also Spektakel Pass).
Sandkraalspoort: off R329 west of Steytlerville.
Sandy's Glen: off R326 off R316 east of Stanford.
Saltpansnek: on R75 southeast of Jansenville (aka Soutpansnek).
Salt River Poort: off N1 south of Nelspoort.
Satansnek: off R56 south of Elliot.
Schoemanspoort: on R328 north of Oudtshoorn (aka Schoeman's Poort; possibly Cango Poort?). 1862
Schurfdeberg Pass: early pass, east of Tulbagh (aka Skurweberg Pass; now Agter Witzenberg Pass. See also Witsenberg Pass). 1780
Sederbergpas: see Cedarberg Pass (aka Cederbergenpas).
Sewefontein Pass: on DR1840.
Seweweeks Poort: off R62 east of Ladismith (aka Seven Weeks Poort). 1862
Shaw's Mountain Pass: on R320 south of Caledon (aka Shaw's Pass – incorrectly!). 1825
Silvermine Pass: see Ou Kaapse Weg. 1811
Sir Lowry's Pass: on N2 east of Somerset West (aka Sir Lowry Cole's Pass initially. Earlier passes: Gantouw, Eland's Pad, Elands Path, Elands Pat, Hottentots Holland Kloof). 1662
Skaapkraalpoort: on R344 north of Tarkastad.
Skilpadhoogte: off R60 north of Bonnievale.
Skitterykloof: off R355 north of Ceres (see also Katbakkiespas; Peerboomskloof).
Skuinshoogte Pass: on R357 northeast of Nieuwoudtville.
Skurweberg Pass: see Schurfdeberg Pass. 1780
Slabberts Pass: off N9 north of Graaff-Reinet, on MR605 (see also Loskop Pass, Rubidge Kloof Pass).

Snydersfonteinpas: on R381 south of Loxton (aka Snydersfonteinhoogte Pass).
Snyderspoort: off R353 north of Sutherland (see also Quaggasfontein Poort).
Soetendalpoort: off R407 west of Willowmore.
Southey Pass: see Tradouw Pass. 1873
Soutpansnek: see Saltpansnek.
Spektakel Pass: on R355 west of Springbok (aka Spektakelbergpas. See also Sandhoogte). 1896
Spreeufonteinpoort: off R337 northeast of Willowmore.
Steenberg Pass: see Ou Kaapse Weg. 1811
Stettynskloof: off N1 southwest of Worcester.
Stoepelshoogte: southwest of Robertson (aka Nuwehoogte?).
Storms River Pass: off N2 east of Plettenberg Bay. 1885
Stormsvlei Pass: somewhere 33/34°S, 23/24°E.
Stormsvleipoort: on R317, off N2, north of Stormsvlei.
Strykhoogte Pass: south of Robertson, east of McGregor.
Strypoort: off R56 southeast of Steynsburg (aka Strydpoort).
Studers Pass: off N7 northeast of Garies. 1933
Suuranys Pass: somewhere 33/34°S, 24/25°E; on DR1800? (see also Moordenaarskloof).
Suurbergpas: on R335 north of Addo (aka Zuurberg Pass; possibly also Addo Heights. See also Doringnek). 1858
Swaarmoedpas: off R46 east of Ceres (on DR1452?).
Swaerhoekpas: on R337 southwest of Cradock (aka Swagershoek Pass. See also Buffelshoek Pass).
Swanepoelspoort: on R337 northeast of Willowmore.
Swartberg Pass: on R328 south of Prince Albert (aka Zwartberg Pass; the Great Zwarte Berg Pass). 1888
Swartnek: on R396 southwest of Barkly East (see also Greylingspas, Killians Pass, Perdenek and Rossouwsberg Pass).
Swart River Pass: on The Passes Road east of George (aka Zwart River Hoogte. See also Hoogekraal, Homtini, Kaaimans, Karatara, Phantom, Touws River Passes). 1871
Swartruggens: 1843
Swempoort: on R392 north of Dordrecht (see also Weenen Pass).

Taandjies Nek: on R63 southwest of Carnarvon. 1969
Table Mountain 'bridle path', from Constantia Nek. 1890s

Tafelberghoogte: on R327 south of Van Wyksdorp.
Tarka Pass: off R337 south of Cradock; on DR2481? (aka Tarka Bothapas).
Teekloof Pass: on R353 south of Fraserburg. 1890s
Telemachus Poort: on N6 north of Jamestown (aka Modderpoort). 1955
Thebles: off R63 northwest of Fort Beaufort.
Theekloof Pass: see Teekloof Pass.
Theewaterskloof: on R43 south of Villiersdorp (aka Floorshoogte).
Theronsberg Pass: on R46 northeast of Ceres (see also Hottentotskloof, Karoopoort). 1994
Thompson's Pass: see Rooihoogtepas. 1877
Thornkloof: on R72 west of Alexandria.
Tierkloof: somewhere 32/33°S, 24/25°E; on DR2393? (see also Nardausberg Pass).
Tierpoort: on R32 northwest of Britstown.
Tiger Kloof: see Messelpad Pass (aka Tierkloof). 1871
Tokai–Hout Bay 'old road'. 1897
Toorwaterpoort: off R341 east of De Rust (aka Toverwater Pass, Towerwaterpoort). 1843
Touws River Pass: on The Passes Road (see also Hoogekraal, Homtini, Kaaimans, Karatara, Phantom, Swart River Passes). 1871
Tradouws Pass: on R324 south of Barrydale (aka Southey Pass; Southeyspas; Tradou Pass). 1873
Trek-aan-Touw: early pass, east of George (aka Krakede-Kau (maidens' ford)). 1812
Trompetterspoort: off R329 northeast of Willowmore (see also Maroegapoort).
Tulbagh Kloof Pass: see Nieuwekloof Pass. 1860

Uitkykberg Pass: on DR2406
Uitkyk Pass: off R364 north of Wuppertal (possibly aka Hoekseberg Pass. See also Bidouw, Hoekseberg, Kleinhoog and Koudeberg Passes).
Uitkykpas: see Cedarberg Pass.
Umzimvubu Pass: on N2.
Uniondale Heights: on N9 north of Uniondale.
Uniondale–Avontuur Pass (replaced by Uniondale Poort). 1867
Uniondale Poort: on R339 south of Uniondale (aka Uniondale Kloof; Uniondale Road; Avontuurkloof). 1940s

Vaalkrantz: on R390 north of Steynsburg.
Vanderkloof Pass: off R48 southwest of Vanderkloof.
Van der Stel Pass: off N2 north of Bot River.
Van der Waltspoort: off R63 southwest of Victoria West. 1994

Van Rhyns Pass: on R27 west of Nieuwoudtville. 1880
Van Ryneveld Pass: under the silt of the Van Ryneveld Dam (aka Van Ryneveldt Pass. See also Oubergpas, Rooikranshoogte, Voetpadhoogte). 1832
Van Stadens Pass: off N2 west of Port Elizabeth. 1867
Varkkloof: off N7 southeast of Clanwilliam (see also Blinkbergpas, Cedarberg Pass, Grootrivierhoogte, Kriedouw Kloof and Nieuwoudtspas).
(Die) Venster: on R46 west of Touws River. 1887
Ventersberg Pass: off R392 north of Dordrecht.
Verkeerde Kloof: see Caledon Kloof. 1807
Verlate Kloof: on R354 southwest of Sutherland (aka Verlaten Kloof. See also Appelfontein Hoogte, Rooikloof). 1877
Versveld Pass: on DR2161 west of Piketberg (aka Versfeld Pass).
Verwatersnek: off R31 north of Olifantshoek.
Victoria Road: on M6, Sea Point to Hout Bay. 1888
Viljoenshoogte: south of Robertson, southeast of McGregor.
Viljoens Pass: on R321 north of Grabouw. 1860s
Vloksberg Pass: down the Roggeveldberge from Sutherland. 1877
Voetpadhoogte: on R63 southeast of Murraysburg (see also Ouberg Pass, Rooikranshoogte, Van Ryneveld Pass).
Voetpadskloof: on R323 north of Riversdale (see also Garcia's Pass, Langkloof, Muiskraal Pass).
Volstruispoort: on R384 southwest of Vosburg.
Vyf Myl Poort: on N7 south of Vioolsdrif. 1965

Waaipoort: on R329 northeast of Steytlerville (aka Waaipoort Pass. See also Noorspoort).
Wagenaar's Kloof: see De Jager's Pass. 1880
Wapadsbergpas: on R61 northwest of Cradock (aka Wagenpads Berg Pass). 1943
Wapadskloof: on R335 southeast of Somerset East.
Wapadsnek: off R29 southeast of Oudtshoorn.
Wasbank Pass: south of Laingsburg, on MR315.
Waterkransberg Pass: on DR2414.
Waterkrans Pass: somewhere 31/32°S, 22/23°E.
Waterkrans Pass: off R63 north of Graaff-Reinet, on MR604 (aka Waterkransberg Pass. See also Osfontein Pass, Leeukloof Pass).
Weenen Pass: on R392 north of Dordrecht (see also Swempoort).
Welgevonden: see Caledon Kloof. 1807
Wepenersnek: on R56 east of Matatiele.
White's Road: down to Wilderness. 1905
Wienands Nek: off R63 northwest of Bedford (aka Wienands Pass).
Wildehondskloof: on R62 west of Barrydale (see also Op de Tradouw, Poortjieskloof Pass).
Wildhondnek: on R390 north of Cradock.
Wildepaardehoek Pass: off N7 southwest of Springbok (aka Wildeperdehoek Pass. See also Messelpad Pass). 1871
Wilderness Pass: see Kaaiman's Gat Pass. 1752
William's Pass: off N7 west of Kamieskroon (see also Grootvlei Pass, Killians Pass).
Windheuwelpoort: on N9 northeast of Willowmore.
Winterhoek Pass: see Carlton Heights(?).
Witelskloof: off N7 southwest of Clanwilliam (on DR2184?).
Withoogte: on R388 north of Richmond.
Witkophoogte: off N6 west of Jamestown (see also Diepkloof Pass, Kapokkraalhoogte).
Witkransnek: on N10 southeast of Middelburg.
Witkranspoort: on R407 east of Prince Albert (see also Kredouws Pass).
Witlieshoogte Pass: on R398 west of Middelburg, on MR610 (see also Lessingshoogte Pass).
Witnekpas: between Nieu-Bethesda and Bethesda Road Station. 1994
Witpoort: on R337 east of Willowmore.
Witsenberg Pass: early pass, east of Tulbagh (aka Witzenberg. See also Schurfdeberg Pass). 1780
Witwaterspoort: off N1 east of Touwsrivier at Konstabel.
Woest Hill Pass: on DR1969.
Wuppertal–Eselbank Pass: on C516 south of Wuppertal (possibly aka Die Poort). 1800s

Xuka Drift: off R61 east of Engcobo.

Zeebraspoort: off R56 northeast of Middelburg.
Zuurberg Pass: see Suurbergpas. 1858
Zwartberg Pass: see Swartberg Pass. 1888

NOTES

CHAPTER 1

1. To compound this confusion, the cartographers have marked as 'Roodezandspas' a kloof about 15 kilometres to the north of the Klein Berg River gorge. This kloof, between the Groot Winterhoekberge and the Saronsberg, was never developed as a pass. One reason: because of the small mountain which blocks its eastern end. And another: because it is out of direction when considering connections to the main settlement in Cape Town. (Nellmapius, 2001: 23–25)
2. Mossop, 1927: 43–46; Nellmapius, 1998; 2001: 3–6, 27–32.
3. Burman, 1963: 49.
4. Burman, 1963: 49–52; 1988: 74; Nellmapius, 2001: 23.
5. Burman, 1963: 50, 52–54; Nellmapius, 2001: 18, 23.
6. Thunberg, 1793.
7. Burchell, 1822.
8. Steedman, 1835.
9. Theal, 1893–1905, vol. 6: 91, 92.
10. Burman, 1963: 49–55; 1988: 74, 75; Mossop, 1927: 51–56, 172; Nellmapius, 1999: 12, 13; 2001: 23–25.
11. Michell, 1836: 169.
12. Burman, 1963: 52, 54, 55; Nellmapius, 1999: 21; 2001: 16–20.
13. Ross, 1998a: 17, 18; Nellmapius, 1999: 21; 2001: 19.
14. Theal, 1893–1905, vol. 24: 402, 422.
15. Theal, 1893–1905, vol. 6: 39, 44.
16. Theal, 1893–1905, vol. 6: 75.

CHAPTER 2

1. E.g. Ross, 1998b.
2. Burman, 1988: 21, 22, 69.
3. Burman, 1963: 73, 74; 1988: 23.
4. Lichtenstein, 1812–1815.
5. Burman, 1963: 74–76.
6. CGH, Reports of the Central Road Board: report dated 23 April 1855.
7. Burman, 1963: 73–77; Storrar, 1984: 39–43.
8. Ross, 1998a: 19, 20.

CHAPTER 3

1. Burrows, 1994: 4–5; Mossop, 1927: 10; 57–69.
2. Heap, 1970: 75.
3. Burrows, 1994: 17, 18; Heap, 1970: 76; Theal, 1893–1905, vol. 8: 433, 434.
4. Burchell, 1822.
5. Government Advertisement dated 5 June 1812, as in Theal, 1893–1905, vol. 8: 433, 434.
6. Cape Province, 1985: 95.
7. Burman, 1963: Chapter 4; 1988: 52–54; Burrows, 1994: 19–23; Cape Province, 1985: 95, 96.
8. Heap, 1970: 84–88.
9. Bunbury, 1848.
10. Burrows, 1994: 21.
11. Cape Province, 1985: 96; Cape PRE, Annual Report 1958/59, Chapter 7; Ross, 1998a: 16, 17.
12. Ross, 1998a: 16, 17.
13. (Designer and Resident Engineer Brian) Dreyer, 1993; Cape PRE, Annual Report 1983/84.

CHAPTER 4

1. Thunberg, 1793.
2. Burman, 1963: 133; 1988: 75, 76; Michell, 1836: 170.
3. Burchell, 1822.
4. Burman, 1963, 127–131; 1988: 76; Michell, 1836: 170.
5. Bozzoli, 1997: 9.
6. Storrar, 1984: 22–27, 52.
7. Mossop, 1927: 179.
8. Bond, 1956: 114, 115; Mossop, 1927: 179.
9. Williamson, 1948.
10. Ross, 1979.
11. Etienne de Villiers in Ross, 1998a: 27, 28.
12. Cape PRE, Annual Report 1992/93.
13. *Die Burger*, 5 December 1998: 8; Ross, 1999a.

CHAPTER 5

1. Burman, 1981: 10–15.
2. Bell-Cross & Venter, 1991: 13; Burman, 1988: 72–74; Thunberg, 1793.
3. Burman, 1981: 17, 137.

4. Bell-Cross & Venter, 1991: 19.
5. *Leen-plase*, originally *leenings plaatsen* when this the oldest system of land tenure was introduced in the Cape in 1714, and sometimes referred to in the writings of those times as 'loan places', were described as a half-hour's walk in every direction from the house. Payment of the annual 'recognite geld' of 24 rixdollars gave the farmer the right to 'leggen en wijden'. The 'Cradock Proclamation' of 6 August 1813 replaced *leen-plase* with perpetual quitrent (Theal, 1893–1905, vol. 9: 203–208). Most importantly for road-makers, it also reserved to the colonial government the right of making and repairing public roads, and raising materials for that purpose on the quitrent premises. This was a most important legal right for those of us who were entrusted with the responsibility of building and maintaining roads – especially gravel roads – in rural areas. It made it possible for us to improve the location of a road without having to resort to expropriation proceedings (the old road reverted to the farmer – or, if he was unlucky, to his neighbour!), and also to quarry gravel for road construction and maintenance purposes with only a courtesy notification to the owner of the ground, again avoiding legal delays. Of course, these rights applied only to quitrent farms, but very, very few rural properties were held under freehold title.
6. Burrows, 1994: 25, 137.
7. Michell, 1836.
8. Burman, 1981: 140–142.
9. Bell-Cross & Venter, 1991: 17; Terblanche, 1993, 2002.

CHAPTER 6

1. Bell-Cross & Venter, 1991: 6, 7; Burman, 1988: 81.
2. Hopkins, 1955: 171; Van Wyk, 1988.
3. Burman, 1981: 19–22.
4. Hopkins, 1955: 171.
5. Bell-Cross & Venter, 1991: 4; Burman, 1981: 40–43.
6. Storrar, 1984: 74–76.
7. Burman, 1981: 42, 43.
8. Cape PRE, Annual Report 1979/80.
9. De Kock in Ross, 1998a: 12, 13.

10. Storrar, 1984: 76, 77.
11. Burman, 1981: 48–50; Van Wyk, 1988.
12. Bell-Cross & Venter, 1991: 10, 11; Storrar, 1984: 77.
13. Bell-Cross & Venter, 1991: 11; Burman, 1981: 50; Terblanche, 1993, 2002.

CHAPTER 7

1. Forbes, 1965: 7–24.
2. Thunberg, 1793.
3. CGH, G25 of 1857.
4. Forbes, 1965: 147, 148.
5. Lichtenstein, 1812.
6. Latrobe, 1818.
7. Tapson, 1973: ix, x.
8. Grill, 1968: 34, 35.
9. Bell-Cross & Venter, 1991: 23–26; Bell-Cross, pers. comm.
10. Barrow, 1801.
11. Storrar, 1984: 58–63.
12. Nimmo, 1976: 128.

CHAPTER 8

1. Parkes & Williams, 1989: 13.
2. Nimmo, 1976: 46–49.
3. Margaret Parkes, pers. comm.
4. Bell-Cross, pers. comm.
5. Gordon-Brown, 1972: 16, 29.
6. Forbes, 1965: 25, 30.
7. Theal, 1893–1905, vol. 9: 83–89.
8. CGH, Reports G25 of 1857 and G7 of 1858.
9. Meyer, 1999: 1–5.
10. Storrar, 1978: 192, 194.
11. Storrar, 1978: 192, 194.
12. Bulpin, 1986: 309.

CHAPTER 9

1. Theal, 1893–1905, vol. 9: 83–89.
2. Michell, 1836: 171.
3. *The South African Almanac and Directory*, 1831.
4. Judd, 1993.
5. Burman, 1963: 143, 144.
6. Alexander, 1984; Judd, 1993.

CHAPTER 10

1. Bell-Cross & Venter, 1991: 17–19; Burman, 1963: 105–112; 1981: 133–137.
2. Steedman, 1835: 323–325.
3. Michell, 1836: 172.
4. Sayers, 1982: Chapter 16.
5. Burman, 1963: 112.
6. Goetze, 1993: 15–17.
7. Burman, 1963: 99, 104–108; 1981: 138, 139; Marincowitz, 1992: 5–9.
8. *George Museum Society Bulletin* 20; Marincowitz, 1992: 15.
9. *Cape Frontier Times*, 1 February 1848; Marincowitz, 1992: 5–7, 9.
10. Grill, 1968: 104 & 122.
11. Burman, 1981: 139.
12. Franklin, 1975: 37.
13. As quoted in Marincowitz, 1992: 5.

CHAPTER 11

1. Burman, 1963: 37.
2. Burman, 1963: 4, 39, 40; 1973: 107, 108; Nellmapius, 1999: 19.
3. Burman, 1963: 6.
4. Theal, 1893–1905, vol. 19.
5. Holloway, 1824: 250–255, in Theal, 1893–1905, vol. 19.
6. AA, 1987: 276; 1991: 330; Burman, 1963: 41; Van Renssen, 1996.
7. Holloway, 1824: 250–255, in Theal, 1893–1905, vol. 19.
8. Theal, 1893–1905, vol. 34: 139–143.
9. Michell, 1836: 171.
10. Burman, 1963: 37–42; Floor, 1985: 84; Van Renssen, 1996.

CHAPTER 12

1. Michell, 1836.
2. Cape PRE, Annual Report, 1985.
3. Burrows, 1994: 30.
4. Burman, 1963: 35.
5. CGH, *Government Gazette*, 22 April 1853.
6. Burrows, 1994: 27.
7. Mossop, 1927: 83.
8. Semple, 1968: 126.
9. Cape PRE, Annual Report, 1958/59, Chapter 7.
10. Burman, 1963: 31–35; Mossop, 1927: 95.
11. Burman, 1973: 69.

CHAPTER 13

1. Burman, 1963: 56.
2. Lister, 1949: 210–213.
3. Cape Archives, CO 6220, sketch.
4. Mackenzie, 1993: 15–21; Van Zyl, 1995: 15–17, 21.
5. CGH, *Government Gazette;* 1851 Year Book.
6. Mackenzie, 1993: 21.
7. Bond, 1956: 114, 115; Burman, 1963: 56–61; Mackenzie, 1993: 13–41; Storrar, 1984: 31–35; Van Zyl, 1995: 15–20.
8. Mossop, 1927: 170–194; also in Smuts & Alberts, 1988: 19–27.
9. Mossop, 1927: 190; Cape PRE, Annual Report, 1959/60.

CHAPTER 14

1. Haak, 1996: 9; Willis, 1994.
2. Haak, 1996: 19; Marincowitz, 1990: 4; Willis, 1994.
3. Goetze, 1993: 138.
4. Goetze, 1993: 32–38.
5. Burman, 1963: 145–147; 1981: 125–128; Marincowitz, 1990: 3–7; Willis, 1994.
6. Du Plessis, 1976; *Lantern*, January 1988: 71, 72.
7. Burman, 1963: 145–147; 1981: 125–128; Goetze, 1993: 32–38; Haak, 1996: 28, 36.
8. Marincowitz, 1990: 10
9. Haak, 1996: 62.
10. De Kock in Ross, 1998a: 11, 12.
11. Haak, 1996: 67; Marincowitz, 1990: 10–13; Ross, 1998a: 11, 12.
12. Cape PRE, Annual Report 1973/74.
13. PAWC advertisement in *Sunday Times Metro*, 5 November 2000: 12.
14. Murray, 2001. See also the article in *Civil Engineering* (October 2001) reporting on the receipt of SAICE Regional and National Awards for Technical Excellence and of a commendation from the SA Association of Consulting Engineers.
15. *Sunday Times*, 15 October 2000, *Cape Metro*: 1.

CHAPTER 15

1. Goetze, 1993: 40.
2. Haak, 1996: 28.
3. Burman, 1963: 147–149; Goetze, 1993: 40, 41; Storrar, 1984: 91.
4. Robinson, 1978: 89, 90.
5. Burman, 1963: 148; Goetze, 1993: 40, 41.

CHAPTER 16

1. Smalberger, 1975: 65, 69; Steenkamp, 1975: 42.
2. Ross, 1998a and 2000.
3. Kotze, 1996: 3.
4. Dickason, 1978: 35, 36.
5. CGH, G34 of 1856; 1865a; 1866.
6. Cornelissen, 1965: 40; Smalberger, 1975: 81–85.
7. CGH, G8 of 1854.
8. CGH, G8 of 1855; G35 of 1856; G36 of 1857.
9. Hall, 1866. His is a masterly report – I am fortunate to have obtained a copy.
10. Fletcher, 1868: 14, 15.
11. Ron Strybis, pers. comm., 1994.
12. I am fortunate in having obtained a copy of Fletcher's survey plan, covering the entire road from Springbok to the Bay.
13. Fletcher, 1868 and 1869.
14. Cornelissen, 1965: 39.
15. Burman, 1969: 230.
16. Hall, 1866: 6, 14.
17. Fletcher, 1870.

CHAPTER 17

1. Burman, 1981: 22.
2. Cape PRE, Annual Report 1953: 20.
3. Storrar, 1984: 43, 74, 77, 78.
4. Burman, 1963: 81; 1981: 24.
5. Burman, 1981: 22–26.
6. Burman, 1981: 34.
7. Burman, 1981: 24.
8. Cape PRE, Annual Report 1953: 20, 21.
9. Burman, 1981: 24–26.
10. Storrar, 1984: 78, 105.

CHAPTER 18

1. Storrar, 1984: 79.
2. Storrar, 1984: 80.
3. Cape PRE, Annual Report 1951: 11.
4. Coyne, 1998.
5. Cape PRE, Annual Report 1951: 11–14.
6. Willis, 1994.

CHAPTER 19

1. Storrar, 1978: 194.
2. Lister, 1960: 19.
3. Burman, 1973: 63–67.
4. Storrar, 1984: 81–91.
5. Burman, 1973: 64, 65.
6. CGH, 1881.
7. Ross, 1998a: 9–10.

CHAPTER 20

1. Bulpin, 1986: 303, 304; Nimmo, 1976: 45.
2. Bell-Cross & Venter, 1991: 23.
3. Nimmo, 1976: 115.
4. Nimmo 1976: 27, 28.
5. Gordon-Brown, 1972: 17, 35; Parkes, 1989: 13, 14, 16; CGH, G25 of 1857.
6. Burman, 1973: 157--165; Margaret Parkes, pers. comm., 2001; Storrar, 1984: 46, 47.
7. Nimmo, 1976: 77, 79.
8. Nimmo, 1976: 26, 32, 65, 78, map on 116.
9. Bulpin, 1986: 295, 296.
10. Burman, 1973: 150.
11. Nimmo, 1976: 78.
12. Storrar, 1984: 60, 61.
13. Bulpin, 1986: 296.
14. Victorin, in Grill, 1968: 100.
15. Burman, 1973: 159; Storrar, 1984: 58--63.
16. Nimmo, 1976: 79.
17. Bulpin, 1986: 297.
18. Burman, 1973: 165.
19. Tapson, 1973: 94.

CHAPTER 21

1. Goetze, 1993: 78–82, 91; Haak, 1996: 21–30; Storrar, 1984: 91.
2. Goetze, 1993: Chapter 5, Appendix 1; Haak, 1996: 29–32; *Prince Albert Friend*, August 1985; Storrar, 1995: 91.
3. Goetze, 1993: Chapter 5; Haak, 1996: 33–35; Storrar, 1984: 93.
4. Burman, 1963: 149–153; Goetze, 1993: 120–143; Haak, 1996: 35–38; Storrar, 1984: 93, 94.

5. Haak, 1996: 73; *Prince Albert Friend*, August 1995; Willis, 1994.

CHAPTER 22

1. Lowe, 1924.
2. Floor, 1985: 12, 13, 18, 32.
3. Aucamp, 1971.
4. Ross, 1998a: 21–23.

CHAPTER 23

1. Goetze, 1993: 13, 14.
2. Burman, 1981: 75, 76.
3. Burman, 1981: 75; Goetze, 1993: 13, 14.
4. Burman, 1981: 127; Goetze, 1993: 137.
5. AA, 1987: 57; 1991: 129.
6. Sunde, 1993.
7. Burman, 1981: 76; Goetze, 1993: 14.
8. Floyd, 1993.
9. Ross, 1998a: 40.
10. Sunde, 1993.
11. Armstrong, 1993.

CHAPTER 24

1. Spilhaus, 1949: 148.
2. Le Vaillant, 1790.
3. Burman, 1962: 119–122.
4. Anon., n.d.; Cape Province, 1960: 57.
5. Jackson: n.d.
6. Storrar, 1984: 96.
7. Storrar, 1984: 95–97.
8. Parry-Davies, 2000.
9. Gasson, 2001: 27, 28, 34, 35.
10. Mick Mountain, pers. comm., 2002; Trueman, 1999.
11. Lawrence Green, as quoted in Cape Province, 1960: 59.
12. Cape PRE, Annual Report 1959/60: 13, 14.
13. Trueman, 1999.
14. Perry, 1922.
15. Anon, n.d.

CHAPTER 25

1. Sayers, 1982: Chapter 17.
2. Ross, 1998a: 36.
3. Ross, 1998a: 36.
4. Baartman, pers. comm., 1998.

5. Ross, 1998a: 36.
6. Kantey & Templer: Report on Buffelspoort.

CHAPTER 26

1. R.J. Fisk in Ross, 1998a: 14, 15.

CHAPTER 27

1. Floor, 1985: 22–25.
2. Burman, 1973: 75, 76.
3. Burman, 1963: 63.
4. Forbes, 1965: 31.
5. Burchell, 1822.
6. Quoted in Burman, 1963: 67.
7. Theal, 1893–1905, vol. 29: 294–297.
8. Theal, 1893–1905, vol. 27: 306–309.
9. Michell, 1836: 171.
10. Cape PRE, Annual Report 1959/60: 14–17.
11. Floor, 1985: 22–24.
12. Floor, 1985: 24.
13. Cape PRE Annual Report, 1959/60: 14–17.
14. *The Civil Engineer in South Africa*, April 1989.
15. In Ross, 1998a: 35.
16. Cape PRE Annual Report, 1951.
17. Williams, 1998.
18. *Cape Times*, 20 June 1997; *Civil Engineering*, Nov./Dec. 1997.

CHAPTER 28

1. Franklin, 1975: 21, 39.
2. Burman, 1973: 139, 140.
3. Grill, 1968: 29, 30.
4. Sayers, 1982: Chapter 17.
5. Franklin, 1975: 39.
6. Burman, 1993: 60–62.
7. Ross, 1998a: 7, 8.

CHAPTER 29

1. Floor, 1985: 25, 26.
2. Cape PRE, Annual Report 1951: 15–17.
3. Baartman, 1992: 2, 3.
4. Baartman, in Ross, 1998a: 29–32.
5. Hoffman, in Ross, 1998a: 32.
6. Rose, 1997.

CHAPTER 30

1. Cape PRE, Annual Report, 1954.
2. Baartman, pers. comm., 1998.
3. E. Sunde in Ross 1998a: 38.
4. Cape PRE, Annual Report, 1979/80.
5. Cullinan, 1992: 88–91.

CHAPTER 31

1. Burman, 1981: 77.
2. Louis Terblanche, pers. comm.
3. Luttig, n.d.
4. Terblanche, in Ross, 1998a: 39.
5. L. Terblanche, pers. comm., May 1998, May 2000 and June 2002: D. Ackermann, pers. comm., July 2002.
6. Goetze, 1993: 177–185; Haak, 1996: 66.
7. Burman, 1981: 76–79.
8. Burman, 1981: 78.
9. L. Terblanche, pers. comm., 10 January 2002.

CHAPTER 32

1. Burman, 1963: 66.
2. H. Belloc, 'The highway and its vehicles', 1926.

BIBLIOGRAPHY

Alexander, Brian H. 1984. Hex River project: a challenge. *Shandbrief* (Ninham Shand staff magazine), December 1984.

Anon. n.d. The Hout Bay Road. MS in the Hout Bay Museum.

Armstrong, M.G.J. 1993. Huis River Pass: 1964–1966. MS (text included in the SAICE database on 'Mountain passes and poorts of the Cape Province', held in the National Library of South Africa (Cape Town) as MSB 953).

Aucamp, Hennie (ed.) 1971. *Op die Stormberge.* Cape Town: Tafelberg.

AA. 1987. *Off the beaten track.* Cape Town: AA.

AA. 1991. *Southern Africa from the highway.* Cape Town: AA.

Baartman, Pieter B. 1992. Construction of the Outeniqua Pass: 1944–1948. MS (text included in the SAICE database on 'Mountain passes and poorts of the Cape Province', held in the NLSA (CT) as MSB 953).

Baartman, Pieter B. 1999. Reminiscences of a padmaker, 5: Kaaimansgat. *Civil Engineering,* April 1999.

Bain, R.F. ('By a descendant'). 1967. Thomas Charles Bain: road engineer extraordinary. *South African Survey Journal,* 66, December 1967: 10–12.

Barrow, Sir John. 1801 & 1804. *An account of travels into the interior of southern Africa in the years 1797 and 1798,* 2 volumes. London: T. Cadell & W. Davies.

Bateman, David Keith. 1992. Mountain passes of the Cape. MS (text included in the SAICE database on 'Mountain passes and poorts of the Cape Province', held in the NLSA (CT) as MSB 953).

Bell-Cross, Graham & Jansie Venter. 1991. *The passes of the Langeberg and Outeniqua Mountains.* Mossel Bay: Bartolomeu Dias Museum.

Bond, John. 1956. *They were South Africans.* Cape Town: Oxford University Press.

Bozzoli, G.R. 1997. *Forging ahead: South Africa's pioneering engineers.* Johannesburg: Witwatersrand University Press.

Bulpin, T.V. 1986. *Discovering South Africa,* 4th edn. Muizenberg: Treasury of Travel

Bunbury, Sir Charles James Fox. 1848. *Journal of a residence at the Cape of Good Hope with excursions into the interior.* London: John Murray.

Burchell, W.J. 1822 & 1824. *Travels in the interior of Southern Africa,* 2 vols. London: Longman, Hurst, Orme & Brown.

Burman, Jose. 1962. *Safe to the sea.* Cape Town: Human & Rousseau.

Burman, Jose. 1963. *So high the road.* Cape Town: Human & Rousseau.

Burman, Jose. 1969. To the end of the line. *SAS-SAR,* April: 230–232.

Burman, Jose. 1973. *Guide to the Garden Route.* Cape Town: Human & Rousseau.

Burman, Jose. 1981. *The Little Karoo.* Cape Town: Human & Rousseau.

Burman, Jose. 1988. *Towards the far horizon: The story of the ox-wagon in South Africa.* Cape Town: Human & Rousseau.

Burman, Jose. 1993. *To horse and away.* Cape Town: Human & Rousseau.

Burrows, Edmund Hartford. 1994. *Overberg odyssey: People, roads and early days.* Swellendam: The author and the Swellendam Trust.

CGH (Cape of Good Hope). 1844–1859. Reports of the Central Road Board of Commissioners of Public Roads in the Cape Colony.

CGH. G8 of 1854: Correspondence on the subject of the discovery of metals in Namaqualand.

CGH. G8 of 1855: Reports of the Surveyor-General Charles D. Bell on the copper fields of Little Namaqualand and of Commander M.S. Nolloth, of HMS 'Frolic', on the bays and harbours of that coast.

CGH. G34 of 1856: Report of the Civil Commissioner of Namaqualand on the state of the roads from the mining districts to the ports of export.

CGH. G35 of 1856: Provisional report &c by Andrew Wyley.

CGH. G25 of 1857: Report of Andrew Geddes Bain.

CGH. G36 of 1857: Report upon the mineral and geological structure of South Namaqualand and the adjoining mineral districts, by Andrew Wyley.

CGH. A40 of 1865: Construction of the road through the Seven Weeks Poort.

CGH. SC8 of 1865: Report of the Select Committee appointed to consider petitions with regard to the construction of lines of main road from Clanwilliam to Springbok, and on the Bill to authorize the Cape Copper Company to construct a line of tramway or railway between Hondeklip Bay and Riethuis.

CGH. 1865a. Spencer F. Innes for the inhabitants of Namaqualand to the Governor of the Colony, dated 18 August 1865.

CGH. 1866. Memorial to the Colonial Secretary, dated 25 August 1866. Cape Town.

CGH. 1881. Report of the Chief Inspector of Public Works.

CGH. A10 of 1885: Report by Inspector T. Bain, P.W.D., upon the construction of Meiring's Poort, and the best means of opening a pass across the Zwartbergen.

CGH. *Government Gazette*, 4 January 1849. Report dated 17th November 1848 by Inspector Henry Fancourt White to the Board of Commissioners of Public Roads.

Cape Province. 1960. *The Administration of the Cape: 1910–1960*. Cape Town: Provincial Administration.

Cape Province. 1985. *Provincial Administration of the Cape of Good Hope: 1910–1985*.

Cape Provincial Roads Engineer (PRE), Annual Report, 1951, 1952, 1953, 1954, 1956 and 1957, 1958/59, 1959/60, 1979/80, 1983/84, 1992/93.

Cape Times. 8 May 1922. Opening of Chapman's Peak Drive.

Cape Times. 20 June 1997. New Du Toit's Kloof toll road opened.

Civil Engineering. November/December 1997: 15. N1 Du Toitskloof.

Civil Engineering. January 1998: 13, 14. N1 Du Toitskloof.

Civil Engineering. October 1998: 18. Outeniqua Pass reconstruction.

Civil Engineering. January 1999: 13, 14. Outeniqua Pass reconstruction.

Civil Engineering. March/April 2000: 22. Chapman's Peak.

Civil Engineering, October 2001: 6–9. Meiringspoort project a remarkable achievement.

Cornelissen, Alwyn. 1965. *Namaqualand copper history*. Privately printed.

Coyne, Patrick. 1998. MS of book.

Cullinan, Patrick. 1992. *Robert Jacob Gordon: 1743–1795: The man and his travels at the Cape*. Cape Town: Struik Winchester.

Dickason, Graham Brian. 1978. *Cornish immigrants to South Africa*. Cape Town: A.A. Balkema.

Dreyer, Brian A.T. 1993. The Sir Lowry's Pass viaduct: 1980–1981. (Text included in SAICE database on 'Mountain passes and poorts of the Cape Province', held in NLSA (CT) as MSB 953.)

Du Plessis, D.B. 1976. History and development of the port of Mossel Bay. *Civil Engineering*, February 1976: 42–44.

Fisk, Ronnie James. 1992. The mountain road from Gordon's Bay to Steenbras Dam. MS (Text included in the SAICE database on 'Mountain passes and poorts of the Cape Province', held in NLSA (CT) as MSB 953.)

Fletcher, P. 1868, 1869, 1870, 1871. Hondeklip and Springbok Road: Namaqualand. Reports of the Chief Inspector of Public Works, Cape of Good Hope, presented to both Houses of Parliament.

Floor, Bernal C. 1985. *The history of National Roads in South Africa*. Pretoria: National Transport Commission.

Floyd, Godfrey. 1993. Huisrivier Pass. MS (text included in the SAICE database on 'Mountain passes and poorts of the Cape Province', held in the NLSA (CT) as MSB 953).

Forbes, V.S. 1965. *Pioneer travellers of South Africa: 1750–1800*. Cape Town: A.A. Balkema.

'F.R.' 1873. Via Tradouw: An account of a journey from Beaufort West to Swellendam. *Cape Monthly Magazine*, 6, 35, May 1873: 308–313.

Franklin, Margaret. 1975. *The story of Great Brak*. Cape Town: C. Struik.

Gasson, C. Barrie. 2001. The story of Boyes Drive. *Kalk Bay Historical Association Bulletin*, 5, March 2001.

George Herald. 31 July 1997. 'A project to be proud of. The pass is open!'

Goetze, Timothy Mark. 1993. *Thomas Bain, road building and the Zwartberg Pass*. M.A. thesis, University of Stellenbosch.

Gordon-Brown, Alfred. 1972. *An artist's journey along the old Cape post road: 1832–1833*. Cape Town: A.A. Balkema.

Grill, J.W. (ed.) 1968. *J.F. Victorin: Travels in the Cape: 1853–1855*. Translated from the 1863 Swedish edition by Jalmar & Ione Rudner. Cape Town: C. Struik.

Haak, Frieda. 1996. *Prince Albert aan die Voet van die Swartberge: Geskiedeniskalender 1762–1995*. Prince Albert: Fransie Pienaar Museum.

Hall, Richard Thomas. 1866. *Report on the roads and system of transport from the Cape Copper Company's mines to the coast of Namaqualand*. Truro: The author.

Heap, Peggy. 1970. *The story of Hottentots Holland: Social history of Somerset West, The Strand, Gordon's Bay and Sir Lowry's Pass over three centuries*. Cape Town: A.A. Balkema.

Hoffman, J.M. 1992. Mountain passes. MS (text included in the SAICE database on 'Mountain passes and poorts of the Cape Province', held in the NLSA (CT) as MSB 953).

Hopkins, H.C. 1955. *Eeufees-Gedenkboek van die Ned. Geref. Kerk Heidelberg (Kaapland): 1855–1955.* Stellenbosch: Pro Ecclesia.

Jackson, Susanne. n.d. Roads to Hout Bay. MS in the Hout Bay Museum, annotated 'From Susanne Jackson's UCT project'.

Jowell, Phyllis (assisted by Adrienne Folb). 1994. *Joe Jowell of Namaqualand: The story of a modern-day pioneer.* Cape Town: Fernwood Press.

Judd, Mike S. 1993. The Hex River Poort and Pass: 1983–1986. MS (text included in the SAICE database on 'Mountain passes and poorts of the Cape Province', held in the NLSA (CT) as MSB 953).

Kantey, Basil A. 1981. George–Oudtshoorn and Huis River Pass. In A.B.A. Brink (ed.), *Engineering geology in Southern Africa*, vol. 2: 119–123. Silverton: Building Publications.

Koch, Retief. 1994. Artery of the Bokkeveld: The saga of a mountain pass. *Light years* (Old Mutual), July 1994: 14, 15.

Kotze, Gert J. 1996. *Die Bou van die Messelpadpas: 6 Februarie 1867 – 31 Maart 1871.* Springbok: Namaqualand Regional Services Council.

Latrobe, Christian Ignatius. 1818. *Journal of a visit to South Africa in 1815 and 1816.* New York: J. Eastburn.

Le Vaillant, François. 1790. *Travels into the interior parts of Africa, by way of the Cape of Good Hope; in the years 1780, 81, 82, 83, 84 and 85*, 2 vols. London: G.G.J. & J. Robinson.

Lichtenstein, Henry. 1812, 1815. *Travels in southern Africa in the years 1803, 1804, 1805 and 1806*, 2 vols. London: Henry Colburn.

Lister, Georgina. 1960. *The reminiscences of Georgina Lister (1860–1954).* Johannesburg: The Africana Museum.

Lister, Margaret Hermina (ed.). 1949. *Journals of Andrew Geddes Bain: Trader, explorer, soldier, road engineer and geologist.* Cape Town: Van Riebeeck Society.

Loubser, M.P. 1992. Piekenierskloof Pas. MS (text included in the database 'Mountain passes and poorts of the Cape Province', held in the NLSA (CT) as MSB 953).

Lowe, Claude A. 1924. *Queenstown Automobile Club motorists' route book.* Queenstown: The Club.

Luttig, P.C. n.d. 'Hell' (Gamkaskloof). MS in the Fransie Pienaar Museum, Prince Albert.

Mackenzie, Keith. 1993. *The story of Bain's Kloof Pass.* Wellington: The Wellington Museum.

Marincowitz, Helena. 1989. *Swartberg Pass: Masterpiece of a brilliant road engineer*, 2nd edn. Oudtshoorn: Bowles.

Marincowitz, Helena. 1990. *Meiringspoort: Scenic gorge in the Swartberg.* Oudtshoorn: Bowles.

Marincowitz, Helena. 1992. *Montagu Pass and other passes over the Outeniqua Mountains.* George: George Museum Society.

Marincowitz, Helena. 1993. *Gamkaskloof: Unique valley in the Swartberg Mountains.* Prince Albert: Fransie Pienaar Museum.

Meyer, Ilse. 1999. *Prince Alfred's Pass: Spectacular and diverse.* Oudtshoorn: The Author.

Michell, Charles Cornwallis. 1836. On the roads and kloofs in the Cape Colony. *Journal of the Royal Geographical Society*, London, 7: 168–174.

Mossop, E.E. 1916. Old Tulbagh Pass and Bain's Kloof. *Motoring*, 1 March 1916: 29–32.

Mossop, E.E. 1927. *Old Cape highways.* Cape Town: Maskew Miller.

Murray, Tony. 2001. Repair of flood damage in Meiringspoort. *Civil Engineering*, March/April 2001: 38.

Nellmapius, John. 1996. *The Harwarden expedition: 1658.* Somerset West: privately published.

Nellmapius, John. 1999. *Mountains barred the way.* Somerset West: Historic Hottentots Holland Association.

Nellmapius, John. 2001. *Passes and people.* Somerset West: privately published.

Nimmo, Arthur. 1976. *The Knysna story.* Cape Town: Juta.

Parkes, Margaret & V.M. Williams. 1989. *Exploring Knysna's historical countryside.* Knysna: Emu Publishers.

Perry, T.W.W. 15 February 1922. Hout Bay–Chapman's Peak Road. Report to the Divisional Council of the Cape.

Reitz, Deneys. 1929. *On commando.* London: Faber & Faber.

Robinson, A.M. Lewin (ed.). 1978. *Selected articles from the Cape Monthly Magazine (New series 1870–1876).* Cape Town: Van Riebeeck Society.

Rose, Cecil. 1997. *The Outeniqua Pass: 1993–1997.* Cape Town: Kantey & Templer.

Ross, Graham L.D. 1979. Low volume concrete road construction in the early fifties in South Africa. *Concrete/Beton*, 15, September 1979: 8–10.

Ross, Graham L.D. 1993a. Anenous Pass. MS

(text included in the SAICE database on 'Mountain passes and poorts of the Cape Province', held in the NLSA (CT) as MSB 953).
Ross, Graham L.D. 1993b. Thomas Bain: Pioneer roadbuilder. *South African Transport*, 25 (288): 49–51; *Civil Engineering*, March 1994: 27–29.
Ross, Graham L.D. 1996a. *Namaqualand: A transportation-related chronology.* Somerset West: privately distributed.
Ross, Graham L.D. 1996b. *Namaqualand: An annotated bibliography.* Somerset West: privately distributed.
Ross, Graham L.D. 1998a. *Reminiscences about Cape mountain passes.* Johannesburg: SAICE Transportation Engineering Division.
Ross, Graham L.D. 1998b. The interactive role of transportation and the economy of Namaqualand. PhD (Transport Studies) dissertation, University of Stellenbosch.
Ross, Graham L.D. 1999a. Michell's Pass in use for 150 years. *Civil Engineering*, February 1999: 29.
Ross, Graham L.D. 1999b. *Mountain passes, roads and transportation in the Cape: A research document*, 2nd edn. Somerset West: privately distributed.
Ross, Graham L.D. 2000. Namaqualand transport history. *South African Transport* (parts 1 and 2) and *Transport World Africa* (parts 3 through 5). Johannesburg: Bolton Publications.
Ross, Graham L.D. 2001a. Cape mountain pass histories: Meiringspoort. *Civil Engineering*, March/April 2001: 37, 38.
Ross, Graham L.D. 2001b. The mountain passes of the Outeniquas. Proceedings: SAICE 2001 Congress, George. Johannesburg: South African Institution of Civil Engineering.
Sayers, Chas O. 1982. *Looking back on George: A medley of musings and memories.* George: Herold Phoenix.
Semple, R. 1968 (reprint). *Walks and sketches at the Cape of Good Hope, to which is subjoined a journey from Cape Town to Plettenberg's Bay in 1801.* Cape Town: A.A. Balkema.
Smalberger, John M. 1975. *A history of copper mining in Namaqualand: 1846–1931.* Cape Town: C. Struik.
Smith, Ray A.F. 1973. National Roads of the past. *The Civil Engineer in South Africa.* Hallmark 9, January 1973: 253–255.
Smuts, Dene & Paul Alberts. 1988. *Die vergete Grootpad deur Ceres en die Bokkeveld.* Johannesburg: The Gallery Press.
South Africa (Republic of): Department of Transport. 1981. *Gateway to the Garden Route.* Pretoria: The Department.
South Africa (Republic of): Department of Transport. 1988. *N1: Du Toitskloof Toll Tunnel.* Pretoria: The Department.
South African Commercial Advertiser and Cape Town News. 17, 22 and 24 September 1853. Opening of Bain's Pass.
Sparrman, Andrew. 1783. *Voyage to the Cape of Good Hope ... into the country of the Hottentots and Caferes.* Perth, U.K.: R. Morrison & Son.
Spilhaus, M. Whiting. 1949. *The first South Africans, and the laws which governed them ...* Cape Town: Juta.
Standard and Mail, 20 January 1872. Opening of the first section of the new road through Cogman's Kloof.
Steedman, Andrew. 1835. *Wanderings and adventures in the interior of Southern Africa*, 2 vols. London: Longman.
Steenkamp, Willem. 1975. *Land of the thirst king.* Cape Town: Howard Timmins.
Storrar, Patricia. 1978. *Portrait of Plettenberg Bay.* Cape Town: Purnell.
Storrar, Patricia & Gunther Komnick. 1984. *A colossus of roads.* Cape Town: Murray and Roberts/Concor.
Sunde, Edward. 1993. Construction of three passes. MS (text included in the SAICE database on 'Mountain passes and poorts of the Cape Province', held in the NLSA (CT) as MSB 953).
Tanner, Andrew. 1993. Great Brak Pass: 1977–1981. MS (text included in the SAICE database on 'Mountain passes and poorts of the Cape Province', held in the NLSA (CT) as MSB 953).
Tapson, Winifred. 1973. *Timber and tides: The story of Knysna and Plettenberg Bay*, 4th edn. Cape Town: Juta.
Terblanche, Louis J. 1993. Bergpasse en Poorte. MS (text included in the SAICE database on 'Mountain passes and poorts of the Cape Province', held in the NLSA (CT) as MSB 953).
Terblanche, Louis J. 1999. L.J.T. se werk by KPA Paaie: Desember 1949–September 1966. MS.
Terblanche, Louis J. 2002. Garcia's and Robinson Passes. MS.
Theal, George McCall. 1893–1905. *Records of the Cape Colony: 1793–1831*, 36 volumes. London: William Clowes & Sons.
Thunberg, Carl Peter. 1793. *Travels in Europe, Africa and Asia made between the years 1770 and 1779*, 3 vols. London: W. Richardson.
Trueman, Denny. 1999. Video on Chapman's Peak Drive and Hout Bay. Copy held at Hout Bay Museum.

Van Renssen, Gerrit. 1996. Die Jan Joubertsgatbrug: Oudste Brug in die Land. *Civil Engineering*, July 1996: 29.

Van Wyk, Frans. 1988. *Riversdal: 1838–1988*. Privately published

Van Zyl, A.P. 1995. Bainskloof: 'n Historiese Oorsig. Thesis for the professional Diploma in Museum Science, University of Stellenbosch.

Whitehead, Marion. 1997. Outeniqua Pass sets a new high. *George Museum Society Bulletin*, 31, December 1997: 1.

Williams, Earl. 1998. Huguenot Tunnel proving its worth. *Fleetwatch*. March 1998: 52–56.

Williamson, H. John M. 1948. The widening and surfacing of Michell's Pass. *Minutes of Proceedings, South African Institution of Civil Engineers*, 46, 1948: 88–118, 147–163, 221–223, 282–298.

Willis, Rosalie. 1994. *Ponder the passes of the Great Karoo, and other regions of the Western Cape Province*. Beaufort West: Central Karoo District Council.

Younghusband, Peter. 1959. Opening up 'The Hell'. *South African Panorama*, August 1959: 18, 19.

INDEX

Page references in italics indicate sketches or photographs.

A
Ackermann, Danie 137, 138–140
Andries Uys Bridge 40
Anenous Pass 183–188, *184, 185, 187*
Armstrong, A.B. 91
Ashton area 108–113
Atherstone, William 97–100
Attaquas Kloof Pass 32–34, 55, 65
Avontuur area 54–59

B
Baartman, Pieter 40, 137–138, 140, 177–180, 187
Bain, Andrew Geddes 7, 10, 25, 26, 46, 81, 125
    Bain's Kloof Pass 81–87, 111
    Duiwelskop Pass 50
    Du Toit's Kloof 168
    Houw Hoek Pass 78–79
    Meiringspoort 89, 90–91
    Messelpad Pass 102–103
    Michell's Pass 24, 25–28
    Prince Alfred's Pass 56
Bain's Kloof Pass 27–28, 81–84, *82, 83, 84, 85,* 86–88, *86,* 168
Bain, Thomas 7, 8, 26, 62, 97, 98
    Cogmans Kloof Pass 109–113
    Duiwelskop Pass 50
    Garcia's Pass 44
    Groot River & Bloukrans Passes 117–122
    Koo Mountain Pass 114–116, *115*
    Meiringspoort 89–90
    Nieuwekloof Pass 4
    Passes Road 51, 52, 55, 125–130
    Piquiniers Kloof Pass 10–11, 13
    Prince Alfred's Pass 56, 58–59
    Robinson Pass 35–36
    Swartberg Pass 132, 134–135
    Tradouw Pass 39–40
    Tulbagh Kloof Pass 5
    Victoria Road 148–149, *149*
Barkly, Sir Henry 40
Barrydale area 38–43
Bateman, D.K. 11, 156
Bell, Charles 84, 103
Blanco 68
Bland's Jetty 91
Bloukrans Bridge *119*, 122
Bloukrans Pass 117–122, *118, 119, 121, 196*

Boesmanskloof Pass 154–155
Boontjies Kraal 75
Borcherds Bridge 87
Borcherds, Petrus 87
Bosluiskloof Pass 98–100, *98*
Boy Retief Bridge 112
bridges
    Andries Uys Bridge 40
    Bloukrans Bridge *119*, 122
    Borcherds Bridge 87
    Boy Retief Bridge *199*, 112
    Darling Bridge 87
    De Waal Bridge 111
    Great Brak River 171–172
    Grey Bridge 28
    Groot River Bridge *118*, 121,122
    Hodges Bridge 109
    Jan Joubert's Gat Bridge 73, 75
    Knysna River timber bridge *126*
    Letty's Bridge 40
    Loftus Bridge 111
    Oudebrug 78
    Palmiet River Bridge 78
    Pilkington Bridge 87
    Silver River Bridge *128*
    Storms River Bridge 122
    White Bridge 28
Buffelspoort 155–158, *156, 157*
Burchell, William 3, 16, 25, 76–77, 165
Burger's Pass 114–116, *115*

C
Caledon Kloof 141–142
Calitzdorp area
    Gamkaskloof Passes 189–194
    Huis River Pass 141–146
Cats Pass 72–73
Ceres area 24–31
Chapman's Peak Drive 147–148, 149–153, *150, 151, 152, 153*
Citrusdal area 8–13
Cogmans Kloof Pass 108–113, *109, 111, 112, 113*
Cole, Sir Galbraith Lowry 17, 19–20, 78
Cole's Pass 77–78
Constantia Nek Road 148
Cradock Kloof Pass 33, 66–68, *66, 70, 71*
Cradock, Sir John 66

D
Darling Bridge 87
De Kock, John 40, 43, 93–94, 119–122, 123

De Rust area 89–95
De Smidt, Adam 7, 55, 90, 96, 99, 125–130
De Villiers, P.A. 75, 154–155, 168, 176–177
De Waal Bridge 111
De Waal, Sir Frederic 150, 151
Duiwelskop Pass 48–53, *49*
Du Plessis, Otto 192, *193*
Du Toit's Kloof Pass 88, 164–169, *165, 166, 167,* 195, *197*

E
Ecca Pass 26
Eland's Path 40
environmental concerns 23, 30, 43, 53, 59, 93, 95, 120, 157, 168, 172–174, 182

F
Fforde, James 132
Fisk, Ronnie 161–163
Fletcher, Patrick 7, 103, 104–107
Franschhoek area 71–75
Franschhoek Pass 17, 71, 73–75, *74, 75,* 132, 134

G
Gamkaskloof 100, 189, 190, 192
Gamkaskloof Passes 189–194, *190, 191, 193,* 195
Gantouw Pass 1, 14, *16*
   see also Hottentots Holland Kloof
Garcia, Maurice 43, *43*, 44
Garcia's Pass 43–44
George area 33
   Cradock Kloof & Montagu Passes 65–70
   Great Brak Pass 170–175
   Kaaimansgat & Duiwelskop Passes 45–53
   Outeniqua Pass 176–182
   Passes Road 51, 52, 55, 123–130
Glenday, Robert 151, 152, 153
Gordon's Bay area 159–163
Great Brak Pass 172–175, *173, 175,* 196, 197
Great Brak River bridge 171–172
Great Brak River Heights 170–172, *171,* 197
Grey Bridge 28
Grey, Sir George 7, 8, 11, 28
Grey's Pass 8, *10,* 11, *12,* 195
Greyton area 154–155
Grier, William 135
Groot River Bridge *118,* 121, 122
Groot River Pass 117–122, *120,* 196
Gymanshoek Pass *see* Plattekloof Pass

H
Hall, Richard T. *see* Hall, Thomas
Hall, Thomas 103–104, 106, 107, 184
Heidelberg area 37–43
Helshoogte 71, 72

Hendy, C. 110
Hex River Pass 63–64, *63*
Hex River Poort 60–63, *61*
history of roadbuilding 5–7
Hodges Bridge 109
Hoffman, J.M. 11, 155, 156, 168–169, 177, 180–181
Holloway, W.C. 73, 74–75
Holmswood Commission 88
Homtini Pass 123, 130, *130*
Hondeklip Bay area 101–107
Hoogekraal River Pass 123, 130
Hottentots Holland Kloof 1, 14–17, *15,* 71
Hottentots Holland Mountains 14–23, 76–80, 159–163
Hout Bay area 147–153
Houw Hoek Pass 16, 76, 77, 78–80, *78, 79*
Hudson's Pass *see* Plattekloof Pass
Huguenot Tunnel 88, 168–169
Huis River Pass 141, 142–146, *143, 144, 145, 146*
Humansdorp 126

J
Jamestown area 136–140
Jan Joubert's Gat Bridge 73, 75

K
Kaaimansgat Pass 45–48, *46, 50, 51, 52*
Kaaimans River Pass 123, 128
Kalkoenkrantz tunnel 110–111, 111–112, *111*
Karatara River Pass 123, 130
Kingna River 108
Kloof Nek 148
Knysna area
   Knysna River 125
   Paardekop & Prince Alfred's Passes 54–59
   Passes Road 123–130
Knysna River 125
Koo Mountain Pass 114–116, *115*

L
Ladismith area
   Buffelsrivierpoort 155–158
   Huis River Pass 141–146
   Seweweekspoort & Bosluiskloof Pass 96–100
Laingsburg area 155–158
Langeberg Mountains 37–44, 108–113
Langkloof Mountains 54–59
Latrobe, Christian 47, 78
Letty's Bridge 40
Lichtenstein, Henry 3–4, 9, 47, 48, 66
Limietberge 81–88
Loftus Bridge 111
Loubser, M.P. 11

M

McGregor area 154–155
Marais, Charl 150, 151
Meiring, Petrus 89, 91
Meiringspoort 89–95, *91*, *92*, *95*, 131, 142
Messelpad Pass 101–107, *103*, *104*
Michell, Charles Cornwallis 7, 17, *17*, 33, 55, 81, 117
    Cats Pass 73
    Cradock Kloof Pass 67
    Du Toit's Kloof 167, 168
    Hex River Poort 62
    Houw Hoek Pass 77, 78
    Michell's Pass 25–28
    Montagu Pass 68–70, 176
    Nieuwekloof Pass 4
    Sir Lowry's Pass 17–19, 21, 34, 197
    Tsitsikamma Road 117
Michell's Pass 24, 25–31, *26*, *27*, *28*, *30*, *31*, 84, 195
Molteno, Sir John 89
Montagu area
    Burger's Pass 114–116
    Cogmans Kloof Pass 108–113
Montagu, John 4, 7, *7*, 25, 27, 68, 69, 81, 84, 96
Montagu Pass 33, 34, 50, 55, *66*, *67*, 68–70, *68*, *69*, *70*, 176, *179*
Mossel Bay area
    Attaquas Kloof & Robinson Pass 32–36
    Great Brak Pass 170–175
Mostertshoek Pass 24–25, 60, 195

N

Namaqualand area 101–107
    Anenous Pass 183–188
    Messelpad Pass 101–107
Napier, Sir George 21
National Roads 7, 11, 20, 52, 62, 79, 88, 93, 119, 122, 123, 124, 137, 141, 164, 168, 169, 172, 177, 197
Nature's Valley area 117–122
Naudesnek-Pot River Pass 137
Nieuwekloof Pass 2–4, *3*, *4*
Nuwekloof Pass 5

O

Olifants Pad 72
Otto du Plessis Road 189
Oudeberg Pass 26
Oudebrug 78
Oudekloof Pass 1–2
Oudtshoorn area
    Attaquas Kloof & Robinson Pass 32–36
    Gamkaskloof Passes 189–194
    Huis River Pass 141–146

Meiringspoort 89–95
Swartberg Pass 131–135
Outeniqua Mountains 32–36, 176–182
Outeniqua Pass 70, 176–182, *179*, *180*

P

Paardekop Pass 55–56, *56*
Paarl area 164–169
Palmiet River Bridge (Oudebrug) 78
Passes Road 51, 52, 55, 123, *124*, 125–130, *126*, *127*, *128*, *129*, *130*
Penhoek Pass 136–140, *138*, *139*
Phantom Pass 123, 125, 130
Piekenierskloof Pass 8, 11–13, 195
Piketberg area 8–13
Pilkington Bridge 87
Pilkington, George 168, 171–172
Piquiniers Kloof Pass 1, 10–11
Plattekloof Pass 37–38
Port Nolloth area 183–188
Prince Albert area 158
    Bosluiskloof Pass 98–100
    Gamkaskloof Passes 189–194
    Swartberg Pass 131–135
Prince Alfred's Pass 35, 56–59, *57*, *58*
Provincial Roads Department 5, 7, 11, 20, 28, 30, 119, 140, 152, 155, 172, 181

Q

Queenstown area 136–140

R

railways
    route through Caledon Kloof 141
    line through Hex River Valley 64
    Houw Hoek railway pass *79*, 80
    Michell's Pass: railway to Ceres 28
    Montagu Pass: railway pass behind George *69*
    Port Nolloth-Okiep railroad 103, 106, 107, 184, *184*
    Sir Lowry's Pass: railway pass *19*, 22
    through Toorwater Poort 90
    Touws River to Ladismith 157
    Tulbagh Kloof Pass 5
Reitz, Deneys 190–191
remskoene 18, 197
Rex, George 123
Riversdale area 43–44
Rivers, Joshua 102
Robinson, M.R. 36, 58, 104, 109
Robinson Pass 33, 34–36, *35*, 176
rock-splitting technique 87, 90–901
Roodezand Passes 1–5, 71
Rooihoogte Pass 114, 115, 116
Rooinek Pass 156

S

Sandhills Cutting 62
Sauer, Paul 181, 186
Schoeman's Poort 134
Schonfeldt, Detlef Siegfried 165, 166
Scott, Dave 20–23
Seweweekspoort 96–98, *97*, *99*, *100*, 131
Silver River Bridge *128*
Silver River Pass 123, 128
Sir Lowry's Pass 17–23, *18*, *19*, *20*, *21*, *23*, 34, 75, 76, 77, 197
Skurweberg Pass 25
Smith, Sir Harry 27
Somerset, Lord Charles 71, 73, 74, 166
Springbok area
    Anenous Pass 183–188
    Messelpad Pass 101–107
Stander, Reuben 177
Stanger, W. 68
Steenbras Mountain Road 159–163, *160*, *161*, *162*, *163*
Stormberg Mountains 136–140
Storms River Bridge 122
Storms River gorge 118, 119, 120
Sunde, Edward 187–188
Swartberg Mountains 89–95, 96–100, 131–135, 155–158, 189–194
Swartberg Pass 93, 131–132, *132*, *133*, 134–135, *134*
Swart River Pass 123, 127
Swellendam area 37–43

T

Tanner, Andrew 173–175
Tassie, Jan 134, 135
Terblanche, Louis 44, 192–193, 194
Thunberg, Carl 2–3, 24, 37–38, 46, 55, 125, 165
Toorwater Poort 89–90
Touw River 128–129
Touw River Pass 123, 128–129
Touws River area 60–64

Tradouw Pass 38–40, *38*, *39*, *41*, *42*, 43
Tsitsikamma area 117–122
Tulbagh area 1–5
Tulbagh Kloof Pass 4, *4*, 5
tunnels
    Bain's Kloof Tunnel 86
    Du Toit's Kloof (Kleigat) Tunnel 168
    Hex River Pass: railway tunnels 64
    Huguenot Tunnel 88, 168–169
    Kalkoenkrantz tunnel 110–111, 111–112, *111*
    Meiringspoort: possible tunnel 93
    Sir Lowry's Pass: possible tunnel 23

V

Van der Stel, Simon 72, 148, 165
Van der Stel, Willem Adriaan 2, 165
Van Kervel, Adriaanus 66, 67
Van Plettenberg, Joachim 46–47, 125
Van Rhyns Pass 197
Van Ryneveldt Pass 26
Victoria Road 148–149, *149*
Victorin, Johan 48, 69–70, 171

W

Wellington area 81–88
White Bridge 28
White, Henry Fancourt 68, 170–171, 176, 197
White, Montagu 51
White's Road 51
Wilderness area
    Kaaimansgat & Duiwelskop Passes 45–53
    Passes Road 123–130
Williamson, John 28, 144, 184–186
Witzenberg Pass 25
Worcester area
    Du Toit's Kloof Pass 164–169
    Hex River Pass & Poort 60–64
Wyley, Andrew 103, 106